Dealing with Difficult People

DEALING WITH DIFFICULT PEOPLE

FOR FIRE AND EMS ORGANIZATIONS

JERRY STREICH

Fire Engineering · BOOKS ·

Copyright © 2026 by
Fire Engineering Books
110 S. Hartford Ave., Suite 200
Tulsa, Oklahoma 74120 USA

800.752.9764
+1.918.831.9421
info@fireengineeringbooks.com
www.FireEngineeringBooks.com

Executive Vice President: Eric Schlett
Acquisitions: David Rhodes, Diane Rothschild
Senior Development Editor: Daniel Edward Petrino
Production Manager: Tony Quinn
Customer Success Manager: Lane Nash
Cover Designer: Brandon Ash
Book Designer: Robert Kern, TIPS Publishing Services, Carrboro, NC

Library of Congress Cataloging-in-Publication Data Available on Request

ISBN 9781593706128

Published in the United States of America

1 2 3 4 5   30 29 28 27 26

You are incredibly fortunate if you have a faithful friend who is always there during your successes and defeats, encouraging you to do more and cautioning you to say less. It took me years to realize how important such people are in my life.

My wife, Lori, has been on life's journey with me for over 33 years. Her beautiful soul and kind heart are known to many. She is a strong leader in all that she does, showing our daughters that anything can be accomplished, no matter your gender or sex, race or ethnicity, age, or experience level. This book—and all I do—is dedicated to my cherished friend.

Lori, I have more respect for you than anyone. Thank you for your continued love and support!

Contents

Foreword xi

Preface xiii

Acknowledgments xv

1 Lessons of Leadership 1

Embracing the Military Experience 2

My Fire Service Focus 5

2 Never Forget That You Are the Chief 7

3 Two Down: The Call That Changed Everything 15

4 The Different Types of Difficult People 19

5 Why People Matter 25

6 Dealing with Difficult 31

Defining Difficult 32

Causes 33

Addressing the Issue 34

The Communication Chain 36

7 Dealing with Entitlement 39

Defining Entitlement 40

Causes 41

Addressing the Issue 43

viii   Contents

8   Dealing with Dangerous 45
Defining Dangerous 46
Causes 47
Addressing the Issue 49

9   Dealing with a Narcissist 51
Defining Narcissism 52
Causes 53
Addressing the Issue 54

10   Dealing with Defamation 57
Defining Defamation 58
The Rise of the Keyboard Commander 59
Causes 59
Addressing the Issue 60

11   Dealing with a Workplace Bully 63
Defining Bullying 65
Causes 66
Addressing the Issue 67

12   Dealing with Disgruntled 71
Defining Disgruntled 72
Causes 73
Addressing the Issue 74

13   Dealing with Small-Minded People 77
Defining Small-Mindedness 78
Causes 79
Addressing the Issue 80

14   Dealing with Deviance 83
Defining a Deviant Person 84
Causes 85
Addressing the Issue 86

15   Dealing with a Gossip 89
Defining Gossip 90
Causes 91
Addressing the Issue 92

16 Generational Differences 95

My Experience 96
Defining Generations 96
Closing Thoughts 104

17 Understanding Body Language 107

Features of Body Language 108
Communication Assimilation 109
Closing Thoughts 113

18 Organizational Impacts 115

19 Personal Impacts 119

Difficult Boss 121
"I Like Your Shoes" 122

20 Putting It All Together 123

The Lessons I Didn't Have 123
The Lessons I Wish I'd Had 124
The Challenge and the Charge 124
The Final Takeaway 125
Closing Thoughts 126

Appendix A   Real-World Scenarios 127

Notes 133
Index 141

Foreword

When I learned Chief Streich was writing a book to help fire service leaders learn how to deal with difficult people, I was both elated and disappointed. First let me explain my disappointment. I served 33 years in the fire service, including 22 years as a fire chief. In an effort to be the most prepared leader I could possibly be, I earned bachelor's and master's degrees in business administration and a doctorate. I completed the National Fire Academy's Executive Fire Officer Program and the University of Maryland's National Fire Service Staff and Command course. Unfortunately, as great as all of these educational opportunities were, none of the curriculum in any of those programs prepared me for dealing with the difficult people I would encounter during my career.

For example, as a fire chief in one fire department, after holding an underperforming and misbehaving employee accountable for his bullying behavior, he wrote this message on a whiteboard outside my office: "Happy Happy Joy Joy. I'm as happy as a fired postal worker with an AK-47!" This threatening behavior went on for several years while all my efforts to hold him accountable were rejected or minimalized by my boss. So I and others had to contain his explosive temper (including a rampage where he ripped a door in the fire station off its hinges).

Although this was the worst employee I had to deal with, he was certainly not the only difficult employee I worked with throughout my career. To date, no definitive resource exists for fire service leaders to learn how to deal with our most difficult members.

So why am I elated? Because now, with this book, we finally have a reference for how to deal with our most difficult members! Chief Streich has been working to improve the fire service for many years, and I cannot think of anyone more qualified to teach us these best practices.

As leaders in the fire service, we dedicate ourselves to hiring, developing, and nurturing professionals who are committed to our organization's mission, vision, and core values. When we have a team of dedicated high-performing professionals, leading a fire department can be very fun and tremendously rewarding. But when we are saddled with one or more extremely disruptive and difficult members, running the fire department can be pure hell. I know—I lived it!

I am blessed to have the opportunity to talk to hundreds of fire service leaders every year, and from those conversations, I know many of them are extremely frustrated with the difficult people in their department. These members are counteracting, if not completely destroying, the joy that should come from being a first responder.

This playbook will teach you skills for how to win the battle with your most difficult members. These are powerful lessons that every fire service leader should know but many of us were never taught. Thank you, Chief Streich, for providing this valuable resource. I am confident it will be helpful to so many!

—Dr. Richard B. Gasaway, fire chief (ret.), president and principal consultant, Situational Awareness Matters and author of *Situational Awareness for Emergency Response.*

Preface

When I stepped into the role of fire chief, I expected challenges—budget battles, staffing shortages, tough calls on the fireground. What I didn't expect was harassment, bullying, threats, rage, suicide, and even murder. Those weren't in the job posting, but they quickly became my reality.

Managing people today is more complex and riskier than ever. Technology fuels misinformation and hostility, personal struggles spill into the workplace, and the culture of the firehouse is not immune. What happens online shows up at roll call—and when it does, some of the people we work with become *difficult*.

My journey as a soldier and, later, as a firefighter taught me discipline, service, and honor. But nothing in my military or fire service education prepared me for an employee standing up and screaming at me in a meeting or a firefighter sitting in my office confessing they were considering suicide. No one taught me how to protect my family when someone I supervised threatened me or how to respond when a firefighter's violence at home turned into tragedy that rocked our organization. I had degrees, certifications, and decades of experience but no mental model for what to do when leadership turned dangerous.

This book is written for anyone who finds themselves in a similar position—specifically for leaders who find themselves unprepared for the human problems that can eclipse the fire, the budget, or the call. I am neither a lawyer nor a therapist. Rather, I am a teacher who experienced firsthand the impact of difficult people and wants to pass along the lessons that could protect your health, your team, and your reputation.

In this book, you'll find scenarios, definitions, and strategies to recognize toxic traits: early narcissism, entitlement, bullying, gossip, small-mindedness, disgruntlement, deviance, and more. You'll also find hard-learned lessons on how stress erodes your leadership if you ignore it and how to take care of yourself while you protect your people.

Professional development means *developing*—expanding your mental models through training and realistic scenarios so that you are never caught off guard. When you can see the problem clearly, you can act decisively. And when you act, you protect the 99% of your team who show up every day to do remarkable things.

If you've ever felt alone, questioning whether you're cut out for leadership because of the behavior of a few, I hope this book reassures you: you are not alone. Leaders everywhere are dealing with the same struggles. The question is not whether these challenges will come; it's whether you'll be prepared when they do.

This book is for everyone who believes that leadership is not the only trait we need. We need to manage those who fail to follow both personal and work rules, who consistently force us to pause or even go backward. Designing the best team comes with consistent accountability and focus on protecting those who want to come to work, feel safe, have fun, and go home. I hope that the basic lessons within this book can help you to pave that path. Stay focused. Stay strong. Stay mindful. The future of the fire service and your health depend on it. And know that you are never alone!

Acknowledgments

No book is ever written alone, and this is no exception.

First and foremost, I thank my family. To my wife, Lori, who has stood by me through long nights, high-stress days, and the weight of this career, you are my anchor. To my daughters, Melanie and Natalie, who reminded me that life is bigger than the firehouse and leadership begins at home, thank you for your patience, love, and support. And to my granddaughter, Camila, you inspire me to want a better world for the next generations.

To my friends and colleagues who believe in me—including my close network of chiefs, Dr. Richard Gasaway, Mark Lynde, Ken Prillaman, and John Piper—thank you for your candor, your courage, and your friendship. Our conversations, tough questions, and shared stories have sharpened me more than any single class or conference ever could.

In addition, I am grateful to my teachers, formal and informal, who instilled in me the values of service, discipline, and perseverance. From the U.S. Army to the fire service, from the classroom to the fireground, each of you gave me tools I needed to stand strong in the face of chaos in an ever-changing world. A special acknowledgment goes to Sergeant First Class Glasgow, my drill sergeant in the U.S. Army. He took a young man and turned him into a person with purpose, discipline, and the understanding that presence matters. He introduced me to the meaning of leadership, along with the burden it carries, and encouraged me to always strive to be the best. His voice and examples have echoed through every stage of my career, reminding me that leadership is not a title, but a responsibility.

I am also grateful to the late Chief Bobby Halton and managing editor Diane Rothschild, who believed in my message over a decade ago and pushed me to write. Their trust has filled conference rooms with leaders and firefighters who were experiencing the same challenges. Now, those lessons are captured here in a book that I hope will help even more people to find strength, clarity, and resilience while dealing with those whose mind-set is stopping them from seeing the beauty in life. To Daniel Edward Petrino, senior development editor at Clarion, your infectious attitude and insight added great structure to this book. And my thanks extend to the entire Clarion team, who have reignited the purpose of providing public safety officials with the best in books, videos, education, and events.

Finally, to the difficult people I have met over the years, thank you, too. The challenges you brought into my life became lessons I could never have learned in any classroom or academy across the globe. By testing me, you forced me to build the mental models, strategies, and resilience that I now share in these pages. In your own way, you shaped the leader I became.

1

Lessons of Leadership

When newborns enter the world, they already have inherent abilities and instincts given to them over millions of years of human evolution. I find it amazing that a baby can immediately suck and swallow milk, grab your finger, hear your voice, and cry. Children seek comfort and safety and can adapt to the environments around them. Everything they learn will come from the people who surround them, the places they visit, and their experiences along the journey. Have you ever considered how one child can speak Spanish while their neighbor, at the same age, can speak English? Or how each child in the family sees the world differently? As they age or change environments, they change too. We become who we surround ourselves with, what we see and listen to, and the environments we live in.

With wisdom developed through age, I understand more clearly my beliefs and biases, how the modern world has evolved, and how leadership was presented to me. Growing up, I lived next door to a fire station staffed with full-time firefighters. I could hear the apparatus doors open and could see the fire station from my bedroom window. I got out of bed at night to watch the trucks with red lights drive past my window. My entire room lit up red, and I could make out the silhouette of firefighters in their seats. I imagined what those firefighters were about to face and figured something big was going to happen. I would often visit with the firefighters at the station on my way home from school. They were always welcoming and kind enough to offer me a cookie or a game of bocce ball. Those firefighters made a positive impression on me as a young man, and there was no doubt in my mind that I would become one too.

A local U.S. Army recruiter stopped by my high school to discuss joining. I learned that the military would be the perfect addition to my nonexistent resume to become a firefighter and help pay for my schooling. I took the placement test and scored high enough to enter as a 91A Combat Medic, which is the army's second-largest military occupational specialty, after the infantry. This was a perfect match for my future plan and provided me with lifelong lessons in managing medical emergencies. (Honestly, I was surprised I scored that high, as I never did that well in school. Too many distractions and interruptions in my home life put me in survival mode. I never owned a dog, but somehow it always ate my assignments!)

With my parents' consent, I signed the papers to join the army at 17 years old, through the Delayed Entry Program. Immediately after graduation, I took a plane for the first time and headed to the training center in Fort Bliss, TX. Once I arrived, I was escorted to a bus and transferred to the barracks, where I met my new teachers. Wearing crisp uniforms and big brown hats, they looked much different from my high school instructors.

I quickly learned that these teachers were not messing around. Their voices sounded like rockets as they shouted rapid-fire lingo to get us to respond. My heart was pounding out of my chest as I thought, *I signed up for this?* They marched us all to the barbershop, and within 30 minutes, we were all looking at each other, scratching our now bald heads. We were issued uniforms, bunks, and assignments to keep our barracks safe and clean.

On the first day, I must have done 200 push-ups. Marching with my peers, as the order came to turn left, some went right; when one person messed up, we all paid for it. This did not make sense then, but it is clear as blue sky to me now. As the days progressed, the sergeants sized us up and moved us around, and we learned more about the drill sergeants assigned to us.

I will never forget Sergeant First Class Glasgow. From the first day I stepped off the bus onto the steaming tarmac, this man filled my ear with motivational jargon. He was tall and thin with a large, brimmed drill instructor hat. His battle dress uniform was as crisp as the morning dew, and I swear you could cut an apple with his uniform creases. The insignia on his uniform told me this guy had been through hell, but his voice could have earned him a Grammy when he called cadence. My favorite song of his was "In the Early Morning Rain"; when I sing it today, it takes me back to marching the streets for hours. There is nothing like being in perfect formation with a group of motivated soldiers singing as loud as possible. The entire training facility knew Company B was coming their way.

Embracing the Military Experience

"Streich, front and center!" *What? That's me!* I ran to the front of the formation as I imagined what I could have done wrong. Sergeant Glasgow told the soldiers I was being selected as the squad leader and would be the liaison between our platoon and the drill sergeants. I was shocked because I never had someone look at me as a leader before. I was still trying to figure out what a leader did.

I was given an armband with two stripes to distinguish the squad leader from the rest of the soldiers. When placing the band on my arm, Sergeant Glasgow stated, "Streich, you are a leader. The troops look up to you. In this Army, leadership is all you need." His statement changed my life! Over the next few months, I worked hard to become a soldier. There was no turning back for me; this was part of my plan to become a firefighter.

The military was an excellent experience for me. From the first day, they put us all on the same level. It did not matter where you were from, how much money you had, or if you had the lowest grade point average in your class. You were in the military at that moment, and your mama was nowhere in sight. We all looked alike, with shaven heads and wearing camouflage uniforms with no brand labels to separate our financial status. From that day forward, our purpose was about what we could do to prove we were worthy of becoming soldiers in the world's best army.

The military has a strict plan for training. They must train everyone who volunteers to serve and prepare them for combat. Think about that for a moment; they must prepare them for *combat*, an unfathomable challenge stressed throughout my initial training. We headed to the White Sands Desert, outside El Paso, TX, at the beginning of August. I remember the drill instructors telling me to ensure that everyone had their equipment prepared for marksman training. Supply issued us outdoor gear and a shelter half. I walked throughout the barracks and confirmed that the troops had their equipment in their rucksacks and would be ready to move by 0600 hours. We were ready.

Each morning when it was time to get up, the soldier on fire watch would turn on the lights. I was always up early, made my bed tight enough that a quarter could bounce off it, and prepared for the day. My job was to be ready for the troops if they needed anything. I would also check in with the drill sergeant to see if there were further instructions. I accounted for everyone and reported to the sergeant. They informed us to move into a large trailer attached to a cattle hauler, and we enjoyed a shoulder-to-shoulder—and sometimes head-to-shoulder—ride into the desert. That day, we qualified with the M16A1 automatic rifle and threw a live hand grenade. The thought of throwing a hand grenade frightened me; before this, my only weapon experience had been shooting birds from the sky.

The day intensified as we offloaded the trailer and headed toward the firing range where several new drill sergeants met us. Again, like Sergeant Glasgow, each looked sharp, crisp, and built like a brick house. As they told us how to use the rifle, the instructors made it clear they were serious about ensuring that everyone understood the day's requirements. They issued us a weapon, required us to perform a functions test, and explained that there was a considerable risk of injury or death if we did not follow their instructions. There was no time for horseplay or lack of attention to detail. Before moving to the range, the instructor told us to keep our rifles always pointed downrange: "If your weapon turns around toward me, you will be shot. *Do not move your weapon toward me!*" That message caught my attention, and I was sure to remind my peers as we walked to the range.

As I lay on the hot sand, preparing my weapon for firing, out of the corner of my eye, I could see the drill sergeants behind me. I could only imagine what they had seen during their years of training soldiers. Given that their actions matched their words, I did not take idly the threat of being shot and made sure that my rifle stayed pointing downrange.

Once we finished qualifying with the M16A1, we marched to the hand grenade range, calling cadence, and singing our favorite song. When we arrived at the grenade range, the same rules applied. The risks were higher, though, so this time, we would have a sergeant with us as we were handed a live hand grenade to toss. The process was easy: The instructor would place a hand grenade into our hand with the actuator against the palm of our hand. Once inside the pit, a five-foot by eight-foot underground bunker, he would pull the pin, and you would throw it using an arched toss toward the target. He pulled my pin. "*Throw, throw, throw!*" he yelled as he tossed me onto the ground like a rag doll. *Boom!* The earth shook.

Stress had tired us out by the end of the day. I was proud to have earned Expert Marksmanship on both the rifle and the grenade—which I managed by following directions and paying attention to the instructor's guidance. Although the instructors were intimidating, there was no room for error. People could die that day, and their careers were on the line. Can you imagine taking a group of people from all levels of society and standing with them when you place a live hand grenade in their hand? That is certainly a different level of elevated risk!

Throughout my initial training, I was the only squad leader to stay in the position throughout the entire cycle. That earned me a Certificate of Achievement. When I received this commendation, I was ordered to the front of the platoon, and the sergeant read the inscription:

> For exceptional achievement and outstanding leadership in the performance of his duties as a basic soldier, while serving with Company B, 2D Battalion. Throughout his training, he was extremely successful in grasping the necessary basic leadership skills and general military subjects needed to effectively assist his platoon in mastering the fundamentals of basic soldiering. His performance was noted repeatedly by his chain of command, and his continuous positive effort aided his

company in attaining outstanding scores on end-of-cycle physical training tests, Garrison skills tests, and basic rifle marksmanship. His achievements reflect great credit upon himself, his unit, and the United States Army.

What an honor! That certificate was the spark that set me on fire to succeed.

As I advanced through my military career, I received multiple awards for leadership and was taught that I needed leadership to run a successful team. The drill sergeants communicated a lot about themselves without saying a word. Their presentation and professionalism spoke loudly and instilled many values remaining with me today.

I was a more mature and responsible man after returning from active-duty training. The U.S. Army offered me more than I had expected toward becoming a better person and eventually a firefighter. When I came off active duty, I served in the Army National Guard. Eventually, I moved up the ranks and was promoted as the medical treatment noncommissioned officer in command of a battalion medical aid station. Along with the division surgeon, I managed combat medics within a field hospital 5 miles behind enemy lines.

While managing people, I read leadership books about what it took to become the best leader, in terms of philosophy and experience. I researched values, morals, and professionalism. All the while—over a long time—I gradually worked to improve my reputation and name.

I returned home as a different man who was no longer lost. I was full of confidence, given my achievements and encouragement from the army staff. It was the first time I was called a leader and given the support I needed to grow as a young man. The people I used to hang out with no longer fit the new world I had in mind for my future. Everything looked different.

My Fire Service Focus

Soon after I arrived home from the army, I began training to become a firefighter and eventually was hired as an on-call firefighter in Roseville, MN. The fire chief was young and full of ambition to manage his fire department well. The community was busy with major highways, a railroad track, high-rises, multistory residential buildings, retail malls, parking ramps, a college, and a bulk fuel storage facility. The place was hopping! After a few years, I was promoted to captain and started to define my leadership skills while working as a company officer and incident commander. Our department was selected as the Minnesota Fire Department of the Year in 1992, which instilled a lot of pride in our department.

Finding a job as a full-time firefighter was a challenge in Minnesota, because most of the state's firefighters were volunteers. While searching for a full-time firefighting job, I served as a special deputy within the Hennepin County Water Patrol Division, and I then became a diver. Ultimately, I was hired as a 9-1-1 dispatcher for the largest county facility in the state. I was promoted to tele-communicator sergeant at the sheriff's office on the middle shift, the day's busiest shift. Within the center, I managed a diverse group of dispatchers who were highly competitive, self-driven, and had a keen sense of urgency. They are truly the unsung heroes of the emergency response system.

After finally landing my first full-time job as a firefighter, I moved up through the ranks quickly—initiating programs, building local respect, and attending as many incidents as possible. Some of the programs I created had a countywide impact that caught the eye of other departments across the state. This connected me with many intelligent and influential people over the years, eventually bringing opportunities to me that I had not even been looking for. I participated in statewide and national committees, was selected for fully funded programs, and received numerous awards. I attribute this success to my passion for the fire service and the fact that it never felt like a job.

As a career firefighter, I could focus solely on my duties and add value by being creative with innovative ideas and concepts. To get to the level of chief, I had to move around and accept change to get what I wanted. I adapted to new cultures and leadership styles while continuing my effort to grow. I was on a vertical learning curve, trying to ensure that I represented the fire service well and could act without hesitation when someone called for help.

Like a newborn, the experiences I witnessed along the way and the challenges placed in front of me made me who I am as a leader. In importance, the fire service is not much different from the military; we fight a different war but with the same purpose, to protect our citizens and the community we serve. What I learned, however, was that I needed more than just leadership, as nothing would prepare me for when I needed help the most.

Chapter 1 Reflection Prompts

1. How were your lessons of leadership formed?
2. Which mentors helped you define the importance of your role?
3. In addition to leadership, what else is needed to build strong relationships and teams?

2
Never Forget That You Are the Chief

In 2008, I received an email asking if I would consider applying to become a fire chief for a local fire district. At the time, I was sitting in a brand-new office at city hall, conducting fire investigations and inspecting new and existing properties while responding to calls and managing incidents. It was my dream come true, even though the fire district was 35 minutes from my home and going to and from work took over an hour. Initially, I put the email on the back burner, because I was not interested in becoming a fire chief, having witnessed the challenges other chiefs faced while managing people and working with elected officials.

On a Friday, just before lunch, I remembered the deadline to apply for the fire chief position. I reread the job description and reviewed the minimum requirements. I have always encouraged people to apply for open positions, so I decided to take my own advice and completed an application, especially given that such opportunities did not come along too often, and I would decide if they offered me the job. Just before closing, I dropped off my application and began to wait.

After a few weeks, I received a letter inviting me to interview for the position. That led me to brainstorm scenarios for each question I thought they would ask. The most challenging question for me was, "Why do you want to become a fire chief?" I don't know, the money? Could I make a difference? Why do I want to become a fire chief?

Finally, I realized I wanted to become a fire chief because it was my next level of growth. I was already advancing in the fire service but would stall over time if I did not try something new. I have learned a lot about myself over the years, including that one of my shortcomings is that I can get bored over time. If you want me to update a spreadsheet every day for years, that will not be my ideal job. (Seriously, that is not my lane; however, I will get it done as part of my duties, even though it is not what I aspire to do every day.) But I am the one to go to if you have a problem and need ideas and a path to change your organization. As an analogy, an excellent race car builder may not be the best choice to drive the car in a race.

After several interviews with multiple boards, I was offered the position and eventually accepted it. On starting my new position, I worked with the retiring fire chief for several weeks. During our time together, I learned much more about the fire district and its operation. I have always been amazed at how little one knows when applying for a new position; the job description and interview do not tell the whole story, so in many cases, you go in hoping for the best.

By the end of the first week, I knew there was a lot of work to do, and I decided I needed to go through the stations and equipment on my own. I looked inside every drawer and cabinet, opened every fire engine door, and walked around each fire station in the district. I was given access to the server and began to look through every file online. I reviewed budgets, previous meeting minutes, and personnel files as I began investigating the organization I had agreed to operate and implementing a plan for the future.

Before walking out the fire station door for the last time in August, the former chief gave me a list of people I would likely have trouble with while working there. These five names were written on a napkin. As he walked toward the door, he looked back and told me, "Never forget you are the chief. In this job, sometimes your conscience gets the best of you." I had no idea at the time how valuable that comment would become. He also told me about a manila envelope in the personnel file cabinet with evidence tape on it and told me not to open this particular file unless it was called as evidence in court.

After I wished the chief well and he walked out the door, I dashed to the file cabinet like I was on the U.S. track and field team. The materials within the manila envelope outlined a criminal investigation into one of the employees on the list. He was accused of forging his name on incident reports after the incident was over to get credit and pay for calls he had not attended. The file also included a history of allegrations and complaints that would concern any leader. As I settled in, I solicited feedback from the staff by creating a 75-question electronic survey. I asked them to be professional but answer the questions truthfully, along with potential solutions to the problems they identified. In return for their honesty, I promised to keep their comments anonymous while sharing the collective responses with the entire team.

The survey participation exceeded my expectations and gave me a lot of great feedback. I used the data as a benchmark to gauge future decisions. One question was whether they were aware of any bullies in the department. The replies to this question heightened my awareness of an issue I was not expecting and related to one of the individuals on my list of troublesome employees.

After reading the survey comments, I met individually with everyone on the team. I started each meeting by introducing myself: who I am, where I live, and why I decided to become a fire chief. I then asked them questions and requested feedback on what we could do to improve our organization. A common denominator emerged: there was one person on the team who people were afraid of. At the time, this person, who I will call "Stan" throughout this book, had 19 years of seniority as a firefighter. People described him as always grumpy and mean; one even stated that when seeing Stan's car at the fire station, he would turn around and return home.

Digging further, I learned that many people were afraid of Stan. They related stories about him yelling at them or poking them in the chest with his finger. When I asked their supervisors if this had been reported, they said yes but it continued. This became a standard answer as I continued to interview the team. Silence became normalized, and eventually Stan was allowed to continue. This is not uncommon, as leaders do not usually understand how to overcome the challenges of removing someone from their team.

When I looked at Stan's personnel file, I expected to see these stories of harassment and bullying, but the opposite was true. His personnel file was like looking inside an abandoned house, full of dust and cobwebs. There was nothing I could use as a leverage to instigate him to change his behavior. Consequently, whenever I heard Stan's name, I bumped his meeting down the list.

Finally, I asked Stan into my office. As in other introductory meetings, I described who I am and why I came to the fire district. I then asked who he was and why he stayed with the fire district. He told me about his "run-ins" with the previous chief and asserted that he had been unfairly treated by him. He alleged that there was favoritism and a lack of flexibility.

I thanked Stan for his feedback and asked if he was open to constructive criticism. He said he was, so I began to tell him what I had learned through the other interviews. I told him that people thought he could be a bully sometimes and scared them when he raised his voice. Although I had not yet witnessed any of this behavior, if it were true, it would not meet the mission, vision, and core values I planned to instill within the organization. I told him I did not know him and was not judging him, but I would not tolerate bullying in the fire district. Then, like a reflex, Stan turned the issue around, saying that others were "targeting" him. At the end of the meeting, it was clear that, to him, the problem was someone else's, not his.

That October, I received a phone call from a neighbor of Stan's, stating he had been driving down his street at high speed when their kids were outside. When Stan returned home and pulled up in his car, the neighbor confronted him and told him to slow down. Then, Stan got out of his car, leaned into him, and claimed he had been responding to an emergency. According to the neighbor, Stan was furious, and he asked me to talk to Stan about his speeding, given his concern that it was an ongoing problem affecting the safety of the kids in the neighborhood.

After that phone call, I received two more complaints about Stan's driving conduct. Stan drove a tan Ford Taurus with a red firefighter plate and a sticker indicating that he was a member of our fire district. (In our state, firefighters can obtain a special license plate that is red and features the word "Firefighter" along with the Maltese Cross. The intent is to help others quickly recognize those who may be driving to the fire station or arriving at an emergency scene.) Thus, whenever Stan was road-raging at others, he was readily identified as a member of my fire district, so I received the aggrieved phone calls.

One day, a complainant stopped by my office. He stated he was in his personal vehicle when Stan sped up behind him on the freeway and began to drive aggressively. When he took a phone call, Stan started to shake his head and tailgated him. Eventually, the complainant pulled into the other lane and allowed Stan to pass. The party knew who Stan was, as he had dealt with past issues involving him. I was happy he had stopped by to tell me about his experience and asked him if he would write a statement. He said, "I do not want to take formal action." This disappointed me, as he took the time to come into my office, yet didn't want to document it.

By now I had several complaints about Stan's driving conduct. Clearly, talking to him was not enough to get him to change his behavior. Our policies outlined the fire district's expectations when we approved the special license plate, so I decided to take formal disciplinary action. I met with Stan, with his station chief in attendance, to discuss these policy violations. At the close of this meeting I provided a letter summarizing the details of the infractions and asked him to sign it to document our conversation. He refused to sign the letter and left upset. At our next encounter, I reviewed our initial discussion about meeting our organization's mission, vision, and core values; through his body language, though, it was apparent that Stan did not care.

During my first 6 months as fire chief, most of my time was spent dealing with personnel issues. (Not only was Stan a problem, but several others on the list needed direction.) I was short-staffed in three stations yet had full-time staff with little work to do because of an outdated organizational system. The global recession was taking a toll on how much time people could give to the fire department, with several divorces, foreclosures, layoffs, and families who needed to move out of the state to find jobs. I established a new structure, eliminating the secretarial position and hiring a fire technician to assist with daily operations and respond to emergency calls. I presented it to the board of directors, who unanimously approved it. This set the pace for a new fire district characterized by honor, pride, and purpose.

On August 26, 2009, I received a call from the police chief. Stan had been arrested following a serious domestic incident. Based on what law enforcement reported at the time, I immediately drove to the fire station and drafted a formal notice placing Stan on administrative leave while I investigated. In the letter, I advised him to stay off all fire department property until I notified him otherwise. I pulled his firefighter gear from his locker and secured it in storage. That was the moment I began learning more about the man who had been creating fear in the department and causing trouble since I started my new role.

When the letter was complete, a police detective delivered it to Stan in jail. (I remembered the chain of custody process from when I worked at the sheriff's office and knew Stan would have to sign out for the letter when he was released from jail.) Later that day, Stan called me after reading the letter, asking which policy allowed me to place him on administrative leave. He was angry! I advised him that if the allegations against him were true, they would go against many of our organization's values, and I would not tolerate his actions. He expressed his opinion about that with a few choice words and told me to "watch my back." Initially, I dismissed this statement as venting, but the more I reflected on Stan's actions over the past year, the more I believed I might have to watch out for his revenge. I was already looking over my shoulder, quickly learning that the role of the fire chief was much different than I was expecting it to be.

The situation was escalating. Since that first day when I had been told I would have trouble with him, I had never expected this level of behavior; before that, I had experienced firefighters getting mad because of a policy change or new equipment being added. I was not prepared to deal with someone who could harm someone or would try to drive someone off the road. I asked myself many times, "Why is this guy still here?" If others knew he was a threat, why was he riding on a rig and helping Ms. Smith with her sore back? To me, this firefighter was the number one threat to our organization and needed to go!

Even though I was operating a multicity fire district, I had little legal support. I met with three city administrators regularly, and I updated them on the status of the fire district and this personnel issue. The communities did not have an attorney specializing in employment law, so they directed me to the insurance carrier's attorney, who offered support for such issues. I contacted them several times and told the attorney that I wanted to move to termination based on the data I had discovered and the escalation of the problem. However, they advised that I should not act, since his personnel file had only a few documented incidents. I disagreed; we were not discussing a simple policy violation here. My employees were afraid of this guy, and after he threatened me, my family was on high alert. This seemed like something you would only see on television!

Ultimately, I chose to terminate Stan for the greater good of our organization. Up to that point, I tried everything I could to prevent this outcome; however, it would come at a cost, which was yet to be seen. While Stan was on administrative leave, I searched for answers to all my questions. You could place me in the middle of a disaster with people lying on the ground and flames shooting through the windows, and I would not even break a sweat. However, this single problem had me up and pacing at night, scared and perplexed because I had not experienced this before. I conjured every comparable incident I could think of to finding a solution. Still, the only logical answer I could produce was to terminate him. Protect the good!

Raging through my mind were the following questions:

- Have I done enough?
- What will he do when he learns of the termination?
- Who can I call for more advice?
- Do I need to protect myself and my family? Should I carry a gun?
- Am I a good leader?
- What will others think?
- Should I have taken this job?
- Where is my support?
- Is this the right thing to do for the right reasons? Could this cost me my job?
- Why is this person still working for the district?
- Why does no one else in power seem to care about the threat he poses? Why did the former chief, police chief, and others not provide me with the details of his criminal record?

This single issue was consuming all the space in my mind. I was tired, disengaged, in threat mode, moody, and scared. Was this what becoming a fire chief was all about? While all this was happening, several department veterans turned their backs on me, because, in their words, I was "picking on the old guys." Still, I could not talk to them about the issue because it was way outside of their scope of duty. They did not know about the violent past of the guy who they protected and had worked with for 20 years. If the fire district was a family, it seemed I was not a part of it.

Stan was placed on administrative leave. No one knew where he was staying. Following his arrest and the immediate legal restrictions that were reported, he could no longer remain at the family's home. I had no idea where he was and if he would come back. After crafting the termination letter and notifying leadership, I sent a certified letter informing him of his termination from the fire district. Once it was dropped into the mail, there was no turning back. For a brief period, I thought the nightmare was over. Little did I know the worst was yet to come.

Chapter 2 Reflection Prompts

1. Have you accepted a job you were not prepared for? How did you work through that?
2. Has there ever been a time when someone at work scared you? How did you deal with that?
3. How do you determine if someone is a threat to your organization? What actions would you take to protect your organization?
4. What policies do you have to support your decisions when someone harasses, threatens, or assaults another?
5. Think about the last time a personnel issue kept you up at night. How did you cope with that? What would you have changed about your response?

3

Two Down: The Call That Changed Everything

After sending Stan the certified letter that formally communicated his termination, I focused on my everyday duties. My stress level decreased, and my worrisome nights became fewer. Since I started this position, the primary issue I have been working on was getting to know the organization and navigating personnel issues, including recruitment, retention, organizational structure, and Stan. Now that the main problem was over (or, to my mind, seemed to be), it was time to look at the fire district's operation and find ways to positively impact it. That is why I took the job.

Thursday, October 1, 2009, was productive as we prepared for new recruits to start their training. We set up the training room and equipment for their arrival the following week. At the end of the day, I met with an employee in my office. My office has two windows, one facing the main road and the other facing the fire station parking lot. During my discussion with the employee, I registered the surprise on his face, and before I could ask, he blurted, "Stan is here!" What? I looked out the window behind me and saw his car go past, headed to the back of the fire station. After months of not being seen, he had come to the fire station. The administrative office door was locked, and after a brief time, he drove past my window again now heading back down the road. We were shocked and could only speculate why he would come to the fire station.

I arrived home late that night, spent time with my family, and went to bed. As I have done for decades, I placed my fire department pager in its charger and prayed for a good night of rest. At 10:47 p.m., my pager alerted, with the 9-1-1 dispatcher's message reading, "Attention fire department, shooting," The male caller stated there had been a shooting at the residence. After hearing the address, I jumped out of bed like a firefighter heading to his first fire. I knew it all too well. It was Stan's.

"Chief 1, en route to the shooting," I replied to dispatch. I hopped into my squad, turned on the lights and sirens, and headed that way. "Do you know who has been shot?" The dispatcher knew only what the caller had reported, that there had been a shooting. My heart and mind were racing as fast as the vehicle I was driving. As I traveled, I could hear the traffic on the police radio and requested the responding fire units to stage away from the incident. I did not want them close to the scene until we were advised by law enforcement to proceed. The firefighters responding did not yet know it involved one of their own.

As I was trying to learn more from dispatch and driving at a high rate of speed, my cell phone rang. I could see it was my wife. "Is he going to come and get us?" I have been with her since I was 22 years old, and I could tell she was scared. I told her everything was going to be OK, but in that moment I realized something I did not want to admit. Whatever was happening that night, it was happening to my family too.

When I arrived, I confirmed what no chief ever wants to confirm. Stan had returned home, killed his wife, dialed 9-1-1, and killed himself. There are details I am not including here because they are not necessary to the leadership lesson, and because there are surviving family members who deserve privacy. The point is not the spectacle. The point is what it did to everyone around him and what it exposed about what happens when deviant behavior becomes normal.

Although I did not realize it en route to the incident, when my wife called, I should have turned around. What was I going to do at the scene that mattered more than protecting my family? Like other incidents in the past, I just kept giving my time to the job. Too often it must have sounded like this: "Honey, there is a tornado warning right now. I'm heading to the station. Go to the basement. There are crackers and a water bottle under the stairs for you and the kids." How ridiculous. It took me a long time to acknowledge the fact that I left my family to fend for themselves while I went to work. I was not fully aware of how this situation affected them too. Throughout the year, my wife heard the stories and built her own mental picture of the guy who had threatened me and caused me to lose sleep.

Despite our worst fears, we could not have foreseen how this would end, and it changed our lives. Contemporaneous public reporting described the incident and highlighted what many in the community already knew, that there had been repeated law enforcement involvement over the years, and that his volatility was not new. The same public reporting quoted law enforcement describing him as "one of the top five most violent or potentially violent individuals" in town.[1] Another report stated he had been the subject of dozens of police calls over time, including many at the home.[2] His wife had told police, "I am scared that the next time he gets mad and hits me that it could be the last time."[3]

If Stan was viewed by law enforcement as that dangerous, why did I not know? I would have expected someone, somewhere, would have done everything they could to ensure he was not responding into the community under our name, but that was not the case.

Although I had witnessed his anger, I would have never imagined it would escalate to this level. I have seen many terrible things in my career, but this situation affected me like no other. I was stressed, tired, and afraid. It limited my ability to come home and relax with my family. It also limited my ability to stay focused on the daily duties of the job, so I stayed at work later because it took longer to finish the tasks that needed my attention.

After returning from the scene, I paged the entire fire department to headquarters. It was early morning, and naturally, people were wondering what was happening. I explained that one of our own was dead, along with his wife. Those most affected worked at the same station as Stan but had no idea about the extent of what had been happening outside the firehouse. The incident made headline news and disclosed information that even I had not known. At that point, I was simply hoping for it all to end.

The organization transitioned into healing mode over the next few months. Firefighters were offered counseling and time to reflect on what had happened. The funerals were held separately, and once we made it through them, I hoped we could all come together and work hard to move on from this nightmare; and boy did we ever. The first district flourished with better recruitment, training, and engagement. At the same time, I wondered how the incident with Stan could have come to a better outcome. The bigger question for me professionally, though, was why I was not trained to handle the personnel issues commonly seen in the workplace. Are the public and private sectors promoting people without training? I have several degrees, certifications, and licenses, but I cannot remember a single class on dealing with difficult and complex personnel issues, let alone simple ones.

In training to become a firefighter, we require new recruits to enter a working fire to gain trust in their equipment and understand fire growth and suppression. Yet, when we train leaders and managers, we do not provide them with scenario-based training to help them to develop mental models to fall back on when they are stressed and challenged. This gap is why I wrote this book.

In my mind, I was prepared to become a chief ever since Sergeant Glasgow told me, "Leadership is all you need." Maybe leadership is not all we need, though; Leadership inspires people toward a vision; management builds the systems, accountability, and structure that make that vision work. However, in this case, what we needed was management. Stan was not responsibly managed. His deviant actions became normalized even though the complaints, fears, and retention issues were as clear as day.

Nearly every month since this incident, which has received a lot of attention, I receive emails and phone calls from people asking what they can do with the difficult person they are managing. By sharing about myself and my experience, I hope this book can help you to become a better manager and leader. The most substantial part of your success is doing what is right for the greater good of your organization.

Chapter 3 Reflection Prompts

1. How do you evaluate your time commitment to work, and how does it affect your family?
2. Do you feel you were professionally trained to manage simple and complex personnel issues as a leader and manager? What additional skills and techniques could you develop further?
3. What policies would you like to have to support your decisions? Who do you have in your organization who would support your decisions?
4. Is there someone you know who exhibits behaviors and qualities like Stan's? What are some ways you could deal with someone like Stan?
5. How does managing people affect your personal health?
6. Do you feel that managing people can lead to posttraumatic stress disorder? Think about how this stress could manifest in your everyday life.

4

The Different Types of Difficult People

There are over a thousand books online on how to deal with difficult people—but not even one specific to the fire service. We are a different group of people within the workforce. Most of us do not work for money or glamour; we come to work to serve others. Having interviewed hundreds of candidates, I can almost always predict their answer to the question, "Why do you want to become a firefighter?" Like a rehearsed line in a choir, the majority say, "To serve my community" or "To do something bigger than me." They don't say they are in it for the money. In fact, most firefighters in the United States work a separate full-time job. They serve because they believe in the importance of our work, recognize the impact they can make by helping others, and appreciate the trust and love firefighters receive from the community.

As a firefighter, you are placed on a pedestal and no longer viewed as you. *You* are the firefighter who lives next door. *You* are the firefighter who came to the elementary school. *You* are the firefighter who helped pull the victim out of a burning building. In this job, it is not uncommon to hear your title more than your name.

After becoming a firefighter, you inherit hundreds of years of tradition and trust by simply putting on the uniform. That comes with a lot of responsibility. We have to live up to these expectations and achievements, so that our legacy is not tarnished.

The military taught me how important it is to carry ourselves appropriately, so that those counting upon us can see we are mentally and physically prepared to serve and honor them. Thus, when someone within the organization continues to threaten the environment that allows firefighters to remain engaged in their duty, we have a problem. After my incident with Stan, I gathered as much information about dealing with difficult people as possible. I sought to learn as much as possible about the topic and how to apply it at the fire station. When searching for definitions of difficult people, none of the descriptions I found matched what I had to deal with. I created my own definitions to achieve a broader perspective and clearly identify such individuals inside the fire station.

My Definition of a Difficult Person

1. One who continuously disrupts the mood of the organization.

2. One who fails to meet or maintain the expectations of the mission, vision, and core values.

3. One whose hostile demeanor changes the daily output of others.

This definition helped me to distinguish those who continuously disrupt the mood of the organization from those who don't. Let's face it: we can all be challenging sometimes, but for most of us, this is temporary. This definition allows for that understanding and focuses on those who create havoc regularly.

There are several types of difficult people. Differentiating between them will help generate solutions for how to handle them. In this book, we will look at scenarios involving each of these types.

Inside fire stations, the following are the 10 common types of difficult people:

1. *Difficult.* Constantly challenges authority and spreads negativity
2. *Entitled.* Feels they are owed something or their length of service rules
3. *Dangerous.* Engages in reckless or unethical behavior and may create harm
4. *Narcissistic.* Disregards teamwork and lacks empathy
5. *Defaming.* Tries to ruin reputations
6. *Bullying.* Uses intimidation to manipulate others
7. *Disgruntled.* Always complains and harms morale
8. *Small-minded.* Resists change and new ideas
9. *Deviant.* Ignores policies and procedures
10. *Gossiping.* Spreads rumors that damage trust

Working with one of these individuals can be a challenging experience for anyone. Prolonged exposure to workplace toxicity has been shown to lead to burnout, anxiety, and even posttraumatic stress disorder (PTSD).[1] Already exposed to high-stress situations, firefighters should not have to battle additional emotional exhaustion within their own fire station. Whether it is a colleague, manager, or someone you are assisting on the job, dealing with a difficult person can be draining.

Daily exposure to one of the 10 types of difficult people can lead to a loss in productivity and potentially adverse physical and mental health effects. Have you ever sat in bed and repeatedly thought about what someone said or did to you? Your brain gets caught in a loop of negative thoughts about past events.

When I was dealing with Stan, at times he was all I could think about. I would lie in bed and, once asleep, play out the latest scenario of what *could* happen in troubled dreams; I would get so worked up that I would wake up with a rapid heart rate. *What the hell is wrong with me?* I wondered. To break out of that mind-set, I had to develop a plan for how to proceed the next day. I think my focus was led by fear—not for my health, even though I did consider it, but for getting fired from my job. During that event, I was on the ladder's top rung without any rails to hang onto, looking down at a label stating, "Warning: Do not stand on the top rung." It was already too late!

Several signs can be used to identify a difficult person. Watch out if you see someone displaying the following behaviors:

- Routinely getting mad and yelling at people
- Ignoring organizational norms, policies, and procedures
- Telling unverified stories and gossip to get a reaction or spread rumors
- Consistently challenging the department leadership's decisions and actions.
- Threatening others with force or reprisals—such as threatening to post something online or go to your boss or the community's leadership
- Using fear to get coworkers to react in their favor
- Regularly talking negatively about others and disparaging their character and reputation
- Using their positional power to force others to do what is typically out of the ordinary
- Questioning every purchase, action, and decision as though they know more than anyone else
- Believing the organization owes them for their knowledge and years of service

---

One key to recognizing a difficult person in the fire service is that the issues they create may be *continuous*. Stan was a problem for a long time, and people knew it. He had become the most significant threat within the organization and needed to be dealt with. At no time during the year when I was working on this issue did I believe I could just let it go. I would have only experienced more retention issues and complaints if Stan were allowed to continue his behavior.

Stan's unchecked behavior illustrates one of the biggest failures in leadership: avoiding conflict. As Paul Lencioni has illustrated through his model of dysfunctional teams, those who fail to address toxic behavior cannot function effectively.[2] The reluctance to confront difficult individuals often stems from fear of backlash or discomfort, but the cost of inaction is far greater. There are times when you cannot give up, even if you're fearful of what the outcome could be. I was going to protect those who came to the fire station to serve; they did not return every shift to be poked in the chest, sworn at, or threatened with harm to their reputations. They hired me to be the chief, and that was what I was going to be!

My path to the fire service via the military helped me to develop a service mind-set. In the U.S. Army, I had watched drill sergeants take ordinary people and form them into warriors. They placed live grenades into the hands of their students and guided them through throwing them across the pit. If the student dropped the hand grenade before throwing it, the student and instructor would likely be killed. They marched the soldiers from morning to night to teach them teamwork, discipline, and honor. Why? Because those drill instructors understand that our freedom is not free. You can see that in their uniforms and their demeanor toward their work. They are in the business of serving the citizens of the United States to the exclusion of everything else.

The same can be said about the fire service. As the name indicates, we are here to serve, and that can be done only with quality people and organizational purpose. To get there, you must develop a culture that does not allow an individual to come onto the floor and continuously disrupt your team's upbeat mood and movement. I don't care how many years you have on the job, in this business, we are consistently evolving, and your years of service, although respected, do not drive this engine. Therefore, I ask why: Why do we allow a single person, a few, or the mob to rule the organization's success by being difficult? Why are those on the floor allowing someone to continuously disrupt the mood of the organization? Why is the leadership not taking a strong stance to change behavior? Finally, how can you ever become the elite team most people *think we are* when more fires are occurring inside the station than outside?

The fire service is built on trust, discipline, and a shared mission to serve. Allowing toxic individuals to erode that foundation is unacceptable. As leaders, we must identify, address, and remove workplace toxicity to protect the integrity of our teams. The underlying question is whether you have the courage to take a stand to protect your community.

Chapter 4 Reflection Prompts

1. Do you know someone who consistently changes the mood of the organization? Has it been an ongoing issue? If so, what has allowed the problem to continue?
2. List three reasons why leaders and managers do not properly manage this type of person.
3. What can you do to protect the good within your organization?
4. How do you define a difficult person? How does your definition align with and differ from the 10 types listed here?

5

Why People Matter

Consider yourself as the person who has been hired to start my community's fire department. For years, we contracted fire services from a neighboring department because our area was primarily rural. Today, though, we are a community of over 20,000 residents (about the seating capacity of Madison Square Garden), with a commercial district, multistory residential occupancies, big-box warehouse facilities, and highway corridors with who-knows-what passing through our town. We need our own fire department now, and you're going to start it. Where do you begin?

First, understand that in most communities, no law requires a fire department to exist. We are a service industry—provided to our community at the discretion of our elected leaders. Once the community council or board decides to initiate its own fire department, someone is needed to head it up, usually referred to as the fire chief. Whoever is appointed to this role must ensure that the department offers the best services possible with the level of support (guidance, budget, equipment) given. My own experience is that most community leaders do not know what level of service they want or need—so quit asking them! Instead, educate them on what is best for the residents living, working, and visiting in your community. At the same time, you must create balance within your organization so that your firefighters can serve while also having a life after work.

To initiate a new department, you must collect a lot of data on the community you are preparing to protect from all hazards. Start by learning the community demographics. Then review the community's long-term development plans to determine how the community is expected to grow, how the demographics may change, and what hazards that growth will bring. Using those data, you can then perform a risk analysis to determine the types of emergency incidents expected in the community. Data collected in state and national fire incident reporting systems can also be used to look for commonalities by occupancy type and age group. In this scenario, you have been using contracted services for years, so request data from the contractors, to determine the types of incidents they have been responding to and if any specific time is busier than the rest of the day. You will quickly see how essential data are for measuring the future and understanding the past.

So far, a lot of work has gone into determining what our community looks like from the perspective of emergency management. Still, we have not talked about staffing at all. Until we fully understand how many members we will need and their essential duties, we will only be guessing. After reviewing our community data, you learn that the most probable (i.e., 51% or greater) fire incident that could occur in your community is a residential structure fire. You will also respond to various medical incidents, both trauma and medical. Knowing this finally gets us closer to determining the number and types of people we need to handle our community's emergencies.

First, though, you must specify the number and types of equipment you will need to do the job, given that the most probable type of fire in our community is a residential structure fire. We can now specify the equipment and personnel needed to extinguish the fire safely. How do we determine that? Research! The National Institute of Standards and Technology and the National Fire Protection Association (NFPA) offer accessible information to determine staffing size and best practices. Once you know the types of incidents that we will respond to, the equipment required for mitigation of those incidents, and the skill level your employees will need, you can then focus on building the foundational policies and procedures of your department.

This scenario is intended to illustrate how important people are in carrying out our daily mission as firefighters. We do not just decide to hire a bunch of people and let them run wild inside the fire stations without purpose and direction. We must manage them and ensure that they have the policies, support, training, and equipment they need to do their job well. Once the number of staff for your fire department is determined, you must assess their skill level to complete the job.

For this scenario, you will need 25 firefighters to establish your new fire department. The city council want you to respond to all smells and bells—in other words, any fire-related incidents and high-priority medical and rescue incidents. You will use this information on purpose to choose the skill level of the employees you need. Based on the directive of the council, you will need *technicians,* or those who specialize in a particular skill or craft. In the work firefighters do, there is little room for error. They enter environments immediately dangerous to life and health, using highly technical equipment that places immense stress on their bodies. They treat humans and act in the best interest of victims who are unconscious or trapped. They use equipment that costs taxpayers millions of dollars. We are not searching for someone to enter data here; we are looking for an athlete with the mind-set to learn and win.

Many firefighters do not give themselves enough credit for what they have learned and know as compared with the general public.

When I teach classes, I like to ask firefighters the following question: "When a citizen of the United States is trapped inside another country during civil unrest and they need to be extracted, who would we call to do that?" The attendees think about it, and soon you hear a variety of answers: Navy SEALs, Army Rangers, CIA [Central Intelligence Agency]. Then I ask them why, and a list of reasons comes my way:

- They are well trained.
- They are disciplined.
- They are likely going to succeed.
- They care about the outcome.
- They understand teamwork.
- They have the support of their superiors.
- They follow policy.

And the list goes on. We send these types of teams because they are highly likely to win even when there is little room for error. They are *elite*, and only a few can do what they do. They do not sit in the back of the room complaining because they are doing ladder training again. They do not complain about a policy change that needs to be implemented for their safety. They follow the rules and understand that they were hired for a specific skill set. The same is true of firefighters.

In August 2010, one such elite team was tested when President Barack Obama ordered the Navy SEALs to conduct a special operation in Pakistan. Their mission was to search for and capture Osama bin Laden, believed to be sheltering inside a compound. The Navy SEALs trained for over a year at a simulator compound like the one they thought bin Laden would be in. When the green light was given to go on with the mission, they had to be ready to win—bringing together their training, experience, and teamwork to capture this high-risk target.

Mark Owen, a Navy SEAL on the mission, explained their challenges in his book *No Easy Day*.[1] Before arriving at the target location, the Black Hawk helicopter they were in experienced a mechanical problem and began falling to the ground, landing on a 12-foot privacy fence. The aircraft came down quickly, landed on its nose, and tilted. To continue his mission, Owen had to jump 6 feet from the helicopter and walk under it with the rotors spinning in the dark. Once the helicopter crash-landed and the SEALs were on the ground, they located the compound and made history. Their accomplishment represents a tremendous example of planning, execution, and bravery. The consensus of my class was correct: They needed support, teamwork, advanced training, and more to be an elite team. By understanding the expectations placed upon them and the commitment they made to the people of the United States, they became one of the most respected teams in the world.

When someone needs assistance in your community, who do they get? That's right: they get *you!* Since the 1960s, the United States has promoted three simple digits to request emergency services: 9-1-1. At any time, for any reason, if someone needs help, they can dial 9-1-1 and a telecommunicator will triage their emergency and dispatch the appropriate units to provide service. Federal, state, and local taxes have poured billions of dollars into this system since its inception. Why? To provide the best services possible to the citizens who live, work, and travel in our great nation.

So, I ask you, who do the people get in your community? Is your team elite? Do you work with people who are committed to being the best? Or do you have people complaining about training, policy, and all else in the back of the room? When you look around, are these the people you want coming to your emergency? I know what the community wants. They want their fire department to be all of the following:

- Well trained
- Disciplined
- Successful
- Caring
- Cohesive as a team
- Supported by their superiors
- Conscientious about following policy

In other words, the community wants and expects the same things as they would from an elite team, and they are not worried about the truck's color or the name on the side of it. They need people who can help them with their problems and be nonjudgmental.

This comparison brings us to the main point: people matter! We must search for people with the skills we need and the ambition to continuously strive to improve. The fire service is always in forward motion. Consider where we have come from; nearly everything we use in the fire service has advanced with technology and research. This means we must stay in tune with the advancements in the tools, policies, and procedures we use to remain safe and effective. If we had never changed, we would be pushing horses back into the fire stations today.

We also need to continuously develop as firefighters. Your pay rate—or even years of service—does not determine if you need to continue to improve your skills. The fact is that you do. When you stop wanting to learn or think you know it all, please consider resigning. While I understand it takes work to stay on top of your game, the safety and well-being of others depends on you. You are part of an elite team, and no one in your community can do what you do. And when a firefighter calls a Mayday and needs help, only another firefighter can help them survive.

When I have an emergency, I want the people who paid attention in class, have clear minds, and trust each other. I want people who understand that we are all different and that each of us brings something to the table regardless of age, gender or sex, race or ethnicity, devotion, and class. We do not train to lose, so why tolerate those who want to?

Chapter 5 Reflection Prompts

1. How closely have you investigated your hiring program to ensure that you hire the right people for your agency? What would you like changed?
2. Do you feel you work with an elite team? What can you do to focus more on the department's role within the community?
3. What changes in someone's mind-set have you seen in service to others that either added value or made a negative impact on the organization?

6

Dealing with Difficult

If you've ever struggled with a difficult person at work, you're not alone. Across industries and workplaces, managing challenging personalities is a universal experience. Since I began my journey spreading this message—for my own therapy, really—I have heard from people across the United States and Canada experiencing very similar issues.

Difficulty dealing with one another is a common thread throughout human evolution. Nevertheless, these challenges can prompt tunnel vision, leaving us feeling alone and targeted. Despite this, let me provide you with encouragement: most of the problems we face in the workplace not only have been experienced before but ultimately were resolved.

Scenario

At one time, Bucky was a good firefighter. People looked up to him and followed his lead on the fireground. Because of this ability to lead, he was promoted to the leadership team to help manage the organization. Over time, though, Bucky's mind-set changed, and his leadership skills diminished as he challenged procedures necessary to move the organization forward. Because he was respected, his voice was heard, and some of his recommendations were implemented. He was even assigned to develop a few policies, which never seemed to get completed.

Bucky's influence grew, but not in a positive way. His team began echoing his dissent, a classic case of *groupthink*—where the desire for harmony overrides critical decision-making.[1] Not only was Bucky continuously negative and unproductive, but his hostile behavior was also infectious. Those around him either chose to stand with him or to bail out and escape from affiliation with his madness. Ultimately, those who stayed in Bucky's sphere would likely all fail if there wasn't any positive intervention. Unless they left the team, some would be demoted, and others, in severe cases, would even be terminated.

> Sadly, I have seen this many times. Good, talented people can become difficult. For whatever reason, the person we saw during the interview may evolve into an organizational risk and team destroyer.

A difficult person in the workplace can be a challenge for everyone involved. Dealing with someone who is difficult to work with—whether a colleague, a manager, or a customer—can be frustrating and draining and may even affect your own productivity. Importantly, everyone has a unique personality and communication style, and not everyone will be easy to work with. Regardless, there are ways to manage and work with difficult people in the workplace to make the experience as smooth as possible.

Defining Difficult

Dealing with difficult people involves recognizing and responding to individuals who continuously disrupt the mood of the organization, negatively influence others, and disregard your department's mission, vision, and core values. This is an intentional practice requiring emotional discipline, clear boundaries, and strategic communication to redirect their behavior, protect organizational health, and keep the team focused on solutions and forward progress.

Common characteristics of difficult people include being:

- *Consistently negative.* Difficult people often find the problem in every solution. They complain about policies, leadership decisions, or team members without offering constructive alternatives. Their negativity drains energy and can set the tone for the entire crew. Arriving to work only to hear consistent negativity gets old. This behavior places the complainer in a negative light, even when highly skilled.
- *Disruptive to team cohesion.* Rather than building unity, difficult individuals create division. They stir up conflict, push their own agenda, or influence others to resist leadership. Over time, their behavior can fracture trust, morale, and their own reputation.
- *Resistant to authority and policy.* A difficult person may believe that rules, policies, or leadership directions don't apply to them. They challenge authority, not to improve outcomes, but to maintain control or resist accountability.
- *Emotionally volatile.* Difficult people often lack control over their emotions, swinging quickly from anger to defensiveness. This volatility

creates a tense environment where coworkers may avoid confrontation to keep the peace. Most will remain silent because they don't want to become victims of the perpetrator.

When these traits are unmanaged, they undermine safety, trust, and effectiveness, creating organizational risk. This erodes the culture from the inside out.

Causes

Difficult people are rarely difficult without reason. Their behavior often stems from a mix of personal and organizational factors that, left unaddressed, manifest in disruptive ways. Understanding the context doesn't excuse bad behavior, but it helps leaders to respond with empathy. Potential causes include:

- *Stress or personal struggles.* Stress brings short tempers or defensiveness into the workplace. Similarly, a firefighter experiencing financial strain, relationship issues, or burnout may act out in frustration. What feels like resistance might actually be emotional exhaustion. Over time, stress can lead to a number of health issues that compound the negative mind-set of a difficult person.[2]
- *Lack of emotional awareness or skills.* Some individuals simply lack the emotional intelligence needed to navigate conflict or express themselves appropriately. Moreover, lack of emotional awareness or interpersonal skills may prevent someone from recognizing how their words, tone, attitude, or actions affect others; consequently, they don't realize or they underestimate the impact that their behavior has on team morale and organizational culture. Low emotional awareness can manifest as poor listening skills, defensiveness, or an inability to manage strong emotions, such as anger or frustration. Individuals with low emotional intelligence often struggle with teamwork, leadership, and conflict resolution, which can contribute to workplace tension and reduced performance.[3] Without training, mentorship, or support, these behaviors can persist unchecked or escalate with stress.
- *Organizational misinformation or rumors.* Unclear communication from leadership can create confusion, resentment, or mistrust. In the absence of facts, rumors and assumptions spread, sometimes causing individuals to react defensively or resist change, based on incorrect information. When someone starts a rumor, word spreads like wildfire. Departments should have a fact-checking process to handle rumors and

misinformation, to mitigate this threat to the team's daily workload and safety.

- *Past workplace trauma.* Someone who has been mistreated, overlooked, or disrespected in the past may carry that trauma into new situations. Their behavior may be defensive, controlling, or abrasive—in a misguided attempt to protect themselves from future harm. For example, an employee who has been repeatedly denied a promotion, overlooked for special assignments, or transferred to less desirable stations may begin to expect mistreatment as the norm. Over time, these experiences erode trust, lower resilience, and create a heightened sensitivity to perceived slights. Unresolved workplace trauma can lead to long-term emotional strain, diminished job satisfaction, and maladaptive coping behaviors, all of which impair individuals' abilities to engage constructively in the organization.[4] Leaders who recognize these patterns can respond with empathy while still setting boundaries that uphold accountability and organizational values.

Addressing the Issue

The first step when working with a difficult person is to understand why they are difficult. Getting to the root cause of the behavior allows you to approach the situation with empathy.

Remember that everyone has their own perspective, and that the person's behavior may not be personal—they don't truly know who you are. (I frequently ask those who reach for advice if the person giving them problems eats at their dinner table. If the answer is no, then maybe don't take it personally.) If you feel that you are doing what is best, then drive on. People have the right to talk about you, as long as they are not threatening, defaming, or harassing you.

The following are strategies for addressing difficult behavior:

- *Set clear boundaries.* Once you understand why the person is difficult, establish clear boundaries by communicating your expectations, stating what is and is not acceptable behavior, and standing firm with those boundaries. For example, suppose that the person is constantly interrupting you during meetings. In that case, calmly and assertively remind them that it's vital for everyone to have a chance to speak. Setting boundaries can help the person to realize that their behavior is unacceptable and keep you in control of the situation.

- *Stay calm and professional.* When working with difficult people, you must remain calm and professional. It's easy to get caught up in their emotions and react in a way that is not productive. However, by staying calm, you can avoid getting into an argument and instead focus on finding a solution to the problem. Again, remember that the person's behavior does not reflect you and that changing them is not your job. If you are confident in yourself, you can move through other people's emotions more successfully.

- *Practice active listening.* Another effective strategy is to use *active listening*—that is, not just hearing the words of others but reflecting them back to gain a true understanding of their perspective. This stems from the work of psychologist Carl Rogers, known for his research on humanistic therapy. He developed a *reflective listening* model, which involves paraphrasing, summarizing, and mirroring emotions to ensure that the speaker feels heard.[5] When someone is difficult to work with, they may feel ignored or misunderstood. Instead of reacting immediately, use active listening; repeat or summarize what they have said to confirm your understanding. For example, if a colleague is frustrated about a new policy, you might respond, "It sounds like you're worried this change will make your job harder. Is that correct?" This approach deescalates tension, demonstrates empathy, and creates space for constructive dialog. Remember, we are all different, so understanding the other's point of view can help to resolve the conflict. Their behavior may not be personal; rather, they may be reacting to stress, frustration, or past workplace experiences. By staying calm and practicing active listening, you can navigate difficult conversations more effectively while maintaining professionalism.

- *Document, document, document.* Documenting a difficult person's behavior is crucial. As my friends in the legal field say, "If it's not documented, it didn't happen." Keep personal notes in a journal. This is a simple yet effective way to track interactions and reflect on situations objectively. Take 10 minutes at the end of the day—or immediately after an incident—to record key details while they are still fresh in your mind. Daily journaling fosters reflection, enhances self-awareness, improves decision-making, and boosts overall workplace performance.[6] This doesn't mean writing a book; instead, the practice of transferring thoughts from your mind onto paper helps you to process your experiences and move forward. A few minutes of reflection each day can make a considerable difference.

- *Maintain a trusted peer network.* Having a network of peers with whom you can share information is essential. However, there are a few rules when establishing such relationships, and the first is confidentiality. I have a few close friends I call to get their opinions and ideas; some are in the fire service, and others are not. When I call them, or they call me, we start by stating "I have a scenario." This conveys that we need support from our peer. When I am called, sometimes I just listen, and other times I ask if they want my opinion; In either case, I am not in their shoes and did not experience what they experienced, so I am giving advice based only on the information they provided me with. It is so easy to assume you have all the information, but unless you conducted the investigation and collected actual evidence, either direct or circumstantial, you are working with hearsay—so be careful!

The Communication Chain

Have you played the telephone game? In a room full of people, let them know you will share detailed information with one of the firefighters in the room. That firefighter will then pass on the information to another firefighter, who then passes it to the next, and so on. Once everyone has received the information, you share the exact details you passed on to the first firefighter. Then, the first recipient tells what they heard and passed on, until finally the last person tells the entire room what they heard. I guarantee that what you stated to the first firefighter will not be the same in the end.

The same distortion occurs with information along communication chains within your organization. I experienced this firsthand when I proposed a policy change to reduce distractions in the fire engine during emergency response. My idea was simple: implement a silent cab rule, allowing only critical communication en route to an incident. The goal was to ensure that firefighters would receive clear instructions without unnecessary chatter; this would reduce distractions so that the driver and officer could remain focused on the details of the incident.

Before I formally announced the policy, word spread. But what people heard wasn't what I had proposed. Instead, a firefighter caught wind of the idea and immediately twisted the narrative: "Who is he to tell us we can't talk in our own rig?" he asked, fanning the flames of frustration among the crew. Suddenly, what was meant to be a safety measure turned into a personal attack on autonomy. Before long, rumors spread through the station like a wind-driven fire—and the message being relayed was exaggerated and completely divorced from the original intent.

That situation reminded me of the communication chain exercise. The message the last person receives is often wildly different form the original. This is a powerful demonstration of how quickly information can become distorted, and that can be incredibly dangerous in a workplace that relies on precision.

In any organization, rumors and misinformation can cause unnecessary conflict. Stop sharing information that you have not verified. Sharing information for the sake of sharing it is gossip, and this creates problems. (For more on gossip, see chapter 15.) Get the facts and encourage the people in your life to work on the same theory. If a friend shares something with you, it should not be rebroadcast unless they say it is OK.

In conclusion, working with a difficult person in the workplace can be challenging. Still, everyone has a unique personality and communication style. By understanding the reason why the person is being difficult, setting clear boundaries, staying calm and professional, practicing active listening, documenting the person's behavior, and seeking help from peers, you can manage the situation as smoothly as possible.

Chapter 6 Reflection Prompts

1. How can you prevent yourself from getting frustrated when talking to the same person about the same issue? What can you do to deepen your understanding of the issue?
2. How do you manage people who talk about you negatively?
3. How can you try to understand another person's perspective?

7

Dealing with Entitlement

Entitlement can erode team cohesion, undermine leadership, and create unsafe conditions. In a profession built on hierarchy, trust, and life-or-death teamwork, even small acts of entitlement can disrupt safety and morale. This chapter provides a definition of entitlement in the workplace, examines its psychological and organizational impacts, and discusses strategies to address it proactively.

> ### Scenario
>
> Marcus has been with the fire department for 15 years. Early in his career, he earned the Rookie of the Year award and excelled in leadership training. Over time, however, Marcus began acting as though he were an officer—giving orders, bypassing directives, and changing daily routines assigned by the battalion chief.
>
> One incident involved a critical water rescue drill near a river dam. The battalion chief's orders were precise, given the dam's dangerous currents. Before arrival, Marcus declared that they would deviate from the approved procedure, dismissing the chief as inexperienced.
>
> Such disregard for proper protocol can endanger not only the operation's success but also the lives of fellow firefighters and civilians. According to NFPA 1500, maintaining a culture of safety is paramount.[1]

Defining Entitlement

Entitlement is the belief that one deserves special privileges or treatment simply because of status or tenure.[2] In the fire service, this may manifest as a senior firefighter ignoring policies and procedures—assuming that their experience (or tenure) overrides the chain of command. Although legal and ethical entitlements (e.g., a safe working environment, fair wages) are justifiable, problematic entitlement occurs when individuals ignore legitimate authority and prioritize personal preferences over the team's needs and safety.

Entitled individuals may feel they are owed simply because of who they are, their accomplishments, or their perceived status in the fire department. In the scenario, Marcus receives a paycheck and benefits in return for the skills we need to serve others. He has departed from those fundamental terms of the employment agreement.

Understanding entitlement can help us to counter its negative impacts. In many cases, entitlement may arise from environmental and psychological factors. People from privileged backgrounds may have grown up with a sense of entitlement and privilege, leading them to expect special treatment; conversely, people who have experienced significant trauma or hardship may develop an entitlement mind-set to cope with their challenges and feelings of powerlessness.[3] You already heard how the U.S. Army manages this: they immediately take you to the quartermaster and change your clothes, shave your head, and work you to the bone to break your privilege and build confidence in those who never had any; soon, everyone is on the same page to learn.

Psychologically, entitlement may be linked to narcissism or an inflated sense of self-importance.[4] (For more on narcissism, see chapter 9.) Individuals with elevated levels of entitlement may be more likely to exhibit narcissistic personality traits—such as grandiosity, lack of empathy, and the tendency to exploit or manipulate others. Have you experienced this? Entitled individuals may need constant validation and attention and have a fragile sense of self-esteem that is easily threatened by criticism or rejection. In the fire service, entitled members may also feel that their time in grade (years of service) drives the organization.

Despite its negative attributes, entitlement is not necessarily permanent or unchangeable. Through awareness, reflection, and targeted intervention, individuals can learn to identify and challenge their entitlement mind-set and develop healthier, more empathetic attitudes toward others. In many cases, this may involve developing a greater sense of empathy, compassion, and gratitude for the experiences and perspectives of others and learning to recognize and appreciate the contributions and achievements of those around us. Although

entitlement can take root in many ways, leaders are not powerless against it
and can shift the culture back toward teamwork and accountability. It will take
time and effort, but it can be done.

Common characteristics of those who feel entitled include:

- *Belief that rules do not apply to them.* Entitled individuals often think
  they are the exception to every policy or directive. They operate as
  though their years of service, title, or reputation give them a pass. This
  mind-set erodes discipline and creates frustration for others who follow
  the rules. In high-risk professions such as the fire service, one person
  deciding they are above protocol not only causes friction but can also
  put lives in danger.
- *Expectation of special treatment.* Entitled people expect privileges
  without earning them. This might show up as demanding the best
  assignments, avoiding unpopular tasks, or assuming that personal
  preferences matter more than the mission. Does this bring anyone to
  your mind? This type of demeanor erodes teamwork and creates
  resentment, because others see the imbalance and lose trust in leadership's fairness.
- *Resistance to accountability.* One of the strongest signs of entitlement is
  dodging responsibility when confronted. Rather than owning mistakes
  or lapses in judgment, individuals blame others, deflect criticism, or
  downplay their behavior. This lack of accountability stunts personal
  growth and undermines organizational trust. When excuses replace
  ownership, performance and credibility both suffer.

Causes

Entitlement is a silent killer of morale, performance, and trust. What starts with
one unchecked attitude left unchallenged can quickly ripple across an entire
department—causing friction between labor and management, weakening relationships with the community, and shifting the focus from service to self-interest.
In truth, nobody is entitled to anything beyond the job they have committed to
doing. We are stewards of public trust, not recipients of automatic praise.

The endgame of entitlement is not just eye-rolling or difficult personalities;
it stalls careers, sacrifices credibility, fractures teams, and in extreme cases,
can lead to the complete breakdown of an organization's culture. The longer
entitlement is ignored, the deeper it becomes embedded into the mind-sets of

those surrounded by it. The following are a few common causes of entitlement:

- *Length of service without accountability.* In many fire departments, seniority is mistakenly equated with superiority. Firefighters who receive praise early in their career but without ongoing feedback or accountability may begin to believe the rules no longer apply to them. Moreover, unchecked admiration can foster an attitude of "I've earned the right," regardless of current performance or behavior. Without active supervision and expectations, tenure can breed entitlement rather than leadership. Without feedback and self-awareness, individuals may develop an inflated perception of their contribution and resist accountability.[5] To counter this, leaders must set clear expectations for those in all roles—firefighter, office, or chief. Focusing on the team, rather than the individual, reinforces that every member contributes equally to the mission. Rank should not provide privilege; it should carry responsibility built on trust, which can quickly be lost when accountability is ignored.
- *Misinterpreted praise and recognition.* Recognition is a key motivator, but when humility and team leadership are not expected, an inflated sense of self-importance can result. This mirrors the cultural shift toward narcissism described earlier, in which individuals interpret praise as validation of superiority, rather than as a prompt for continued growth and responsibility. Leaders can disrupt this pattern by creating an environment where positive reinforcement is balanced and shared across the team. Instead of highlighting a single senior member, acknowledge the collective accomplishments of the group. This can take the form of simple acts, such as writing a thank-you card, recognizing someone in front of their peers, or affirming respect for those in higher positions. Praise—not ego—ignites unity.
- *Cultural reinforcement within the department.* Entitlement is not only a personal flaw but also a reflection of environment. In group settings, behavior that is not challenged is often mimicked and spreads. If peers and leaders tolerate behavior that prioritizes ego over teamwork, entitlement becomes normalized. This aligns with social learning theory, which suggests that individuals adopt behaviors modeled by influential peers or authority figures.[6] The antidote to cultural entitlement is open, consistent communication. Teams that take time to meet without distractions, to clarify roles, and to make known their expectations are far less likely to normalize ego-driven behavior. When communication

is strong, entitlement has little room to grow because assumptions are replaced with understanding, and silence is replaced with accountability.

Addressing the Issue

To address entitlement, consistent effort is required. Following a stepwise approach yields the best results. The following are strategies for addressing entitled employees:

- *Encourage empathy and humility.* Leaders can reduce the prevalence of entitlement by reinforcing empathy and humility as core values. Remind employees that no one is above accountability and that the job is about service, not status. Simple acts—such as recognizing team contributions, highlighting shared success, and modeling humility yourself—go a long way in setting the tone. Entitled people thrive on attention; redirect that energy toward appreciation for the team and the mission.
- *Set clear expectations and accountability.* Entitlement often grows when rules are applied inconsistently or feedback is absent. Make sure that your team knows exactly what is expected of them and that policies apply to everyone, regardless of rank or tenure. Tie recognition to performance and teamwork, not seniority or personality. When people know the standards, it is harder for entitlement to take root.
- *Balance praise with responsibility.* Praise and recognition are essential motivators, but they should always be paired with responsibility. When individuals are celebrated without being reminded of their duty to the group, they may begin to believe themselves to be above the team. Leaders should intentionally recognize accomplishments while reinforcing expectations for continued growth, humility, and accountability.
- *Promote a culture of service over status.* We are the fire service, yet entitlement flips that purpose—service—on its head, shifting the focus from serving the public to serving personal interests. Encourage a mind-set where the mission comes first. Talk openly about the dangers of entitlement; reinforce that privileges are earned, not assumed; and make service to others the standard of success.
- *Lead by example.* Nothing kills entitlement faster than a leader who models humility, accountability, and service. When your firefighters see you following the same policies, taking on tough jobs, and treating everyone with respect, they will know that no one is above the mission.

Entitlement fades when culture values effort, accountability, and shared responsibility over ego and privilege.

Entitlement can destroy morale, fracture relationships, and even cost communities their fire service. But by setting expectations, balancing recognition, and reinforcing service over status, leaders can prevent entitlement from taking root. The result will be a culture of humility, accountability, and respect—one where everyone remembers why they signed up in the first place.

Chapter 7 Reflection Prompts

1. Do you know someone within your community who acts entitled? How does it affect those around them? How can you get to the root of an entitled attitude?
2. What can you do to ensure that you are not the one displaying entitlement?
3. If someone on your team has an entitled mind-set, how might you help them to break out of it?
4. How can you show others on your team that entitlement has no place in the fire service?

8
Dealing with Dangerous

Working in the fire service involves high levels of stress and close-knit teamwork. Yet, alongside the camaraderie, there can be individuals whose behaviors pose risks to their colleagues. This chapter explores one such scenario, again involving Stan, whose bullying and violent tendencies highlight the need for firm policies, open communication, and effective mental health resources. By recognizing the warning signs and employing proactive measures, departments can protect personnel while supporting those who need help.

Scenario

Stan was a bully at work, frequently intimidating colleagues and dismissing concerns about his behavior. Initially, no one considered him dangerous—until reports of escalating aggression both at work and in his home life were discovered, which had been filed away, without action. At the time, no one imagined he could be violent at work or hurt someone. However, his continued bullying suggested that he could pose a real threat.

Firefighters often bond under high-pressure conditions. This camaraderie makes it hard not to give a coworker the benefit of the doubt and assume he would never become violent. Yet, no department is immune to the possibility of including a dangerous individual. As the National Institute of Occupational Safety and Health (NIOSH) suggests, identifying early red flags can prevent escalation.[1] Furthermore, recognizing warning signs of escalation can prevent a dangerous person from causing harm.

Defining Dangerous

Understanding the characteristics and motivations of dangerous individuals is essential to develop effective strategies for managing and addressing their behavior.

Common characteristics of dangerous individuals include:

- *Lack of empathy.* Dangerous individuals may show a lack of empathy or concern for the feelings, needs, or rights of others. They may be insensitive to the suffering or distress of those around them and show little remorse or guilt for their harmful actions. This may indicate a lack of empathy, a key trait linked to potential aggression.[2] Do you know someone who may witness tragedy but experience no emotion? Do you hear people say, "Why are we going to those stupid calls?" or "What a crybaby"? Making unempathetic comments can be a sign of a potential problem.
- *Aggression or violence.* Dangerous individuals may exhibit aggressive or violent behavior toward themselves or others. This could include physical or verbal abuse, threats of harm or violence, or a pattern of impulsive or unpredictable behavior. If at any time you feel scared or your gut tells you to call the police, *then do it!* We must not take threats, comments of violence, or aggressive actions like this lightly. There have been too many predictable incidents, and as we say in fire prevention, if it is predictable, it is preventable. Tolerating any type of violence inside the firehouse is unacceptable. Such behavior destroys both individual and organizational trust, and it can take decades to recover. You're at work!
- *Manipulative behavior.* Dangerous individuals may be skilled at manipulating or exploiting others to achieve their own goals. They may use charm, flattery, or deception to gain trust and influence those around them for personal gain. We must watch out for others who may be afraid to say something, to ensure that they know they are not alone.
- *Lack of accountability.* Dangerous individuals may resist taking responsibility for their actions or may deny or minimize the harm they have caused. They may blame others for their behavior or make excuses to avoid consequences. Stan, for instance, responded to accusations by blaming his victims, indicating that he may not recognize or care about the severity of his conduct.

Causes

What actually made Stan do what he did? Even today, I wonder. Understanding why someone like Stan might become dangerous requires a deeper look into human behavior. People are not born dangerous; rather, certain experiences, conditions, and environments can shape harmful tendencies over time. Although there are likely many more causes that can drive someone to harm another, the following list is a compilation of those that I ruminated on over the years during and after the incident with Stan:

- *Childhood trauma and abuse.* Experiencing abuse, neglect, or household dysfunction in childhood can increase risk of violent behavior later in life. The Centers for Disease Control and Prevention have linked early negative experiences, or *adverse childhood experiences*, to a host of behavioral and mental health problems, including aggression, substance abuse, and difficulty regulating emotions.[3] Stan might have learned early in life that control and intimidation were how problems were solved, and if no one taught him otherwise, those behaviors could have followed him into adulthood. This makes sense, as our formative years shape our future.

- *Low self-esteem and insecurity.* Individuals who feel consistently insecure or inferior may act out in aggressive ways to assert dominance or mask their internal struggles. They may bully others to feel powerful or avoid being seen as weak. Stan was a large man, and his presence alone commanded respect; however, he may have been managing internal doubt and lack of self-confidence. Acting out may be a form of self-protection. We all have self-doubt from time to time, but most of us do not harm others or create chaos. This leads me to suspect that inside Stan's mind, he may have been fighting a mental battle with doubt, loss of control, and anger.

- *Lack of emotional regulation.* People who lack coping skills or emotional regulation are more likely to lash out in anger or frustration. This is especially concerning in high-stress environments such as the fire service. Poor impulse control is a key component in many cases of workplace violence. If someone can't handle confrontation without exploding, their behavior isn't just frustrating; it's a threat. Poor emotion regulation and impulse control have been shown to amplify the effect that workplace aggression has on both the victim's and the aggressor's stress levels.[4] For example, when people don't handle their emotions well or lose control in the moment, workplace conflict takes a heavier toll on everyone involved. Both the person lashing out and the person on the receiving end feel more stressed and emotionally drained.

- *Mental health conditions.* Mental health in the fire service is discussed much more now than it was in 1989, when I began my career. At that time, the prevalent attitude was simply, "Suck it up," so we didn't talk about mental health for fear of being labeled weak. Back in the day, talking about what you were feeling wasn't just discouraged—it was dangerous to your reputation. If you admitted you were struggling with stress, anxiety, and trauma, people didn't see it as being human; they saw it as being soft. You could be labeled as someone who couldn't "hack it," or worse, someone who might freeze up on the fireground. That kind of judgment could cost you the respect of your crew, your chances for promotion, or even your spot on the rig. Back then, the culture valued toughness above all else. The problem is, the emotions we all may have do not disappear; they build up, and over time, they take their toll on you, your family, and the people you lead.

- Several mental health disorders, including antisocial personality disorder, borderline personality disorder, PTSD, or bipolar disorder, may contribute to dangerous or erratic behavior—especially when combined with environmental stressors.[5] Importantly, not all mental health illnesses lead to violence. In fact, most people with mental health disorders are not dangerous, but under the right circumstances, untreated or poorly managed psychological conditions can be risk factors.

- *Fear, isolation, and shame.* Dangerous behavior can also stem from deep internal fear of being exposed, being rejected, or losing control. Isolation and shame may feed bitterness and paranoia. When someone feels backed into a corner, their behavior can become more extreme. There was a lot that our fire district did not know about Stan's personal and home life. We spend a lot of time with people at work and really know very little about them. As time went on, I learned a lot about Stan that others did not know and that I couldn't share.

- *Cultural or organizational enablers.* As a consultant, I evaluate other organizations and see how their culture could lead someone to act violently; in particular, when leaders normalize hazing, minimize complaints, and celebrate toxic behaviors, this can feed dangerous behavior. When a dangerous person is tolerated for years, that person and others they work with learn that such behavior is acceptable. Poking someone in the chest and calling them names was as normal as pulling a preconnected handline when I joined Stan's department. Dangerous behavior grows like a weed with thorns on it and is often the result of poor policies, excuses, and weak leadership.

Addressing the Issue

Dealing with dangerous individuals can be challenging and scary. A careful balance is required, protecting yourself and others from harm while respecting the individual's rights and needs.

Strategies for managing and addressing dangerous behavior include:

- *Set clear boundaries.* Establishing clear boundaries and limits on behavior may prevent dangerous individuals from crossing the line into harmful or threatening behavior. A respectful workplace policy is a must. Be sure to mention dangerous behaviors and your actions if they occur. Also, talk about weapons within the fire station: Can I carry a gun while in my personal protective equipment? Where do I put my gun if I come to the station with it? The rules regarding weapons inside the fire station vary by state and locality, but the discussion needs to be had, and rules should be implemented to outline how weapons can or cannot be part of the workplace.
- *Seek professional help.* Dangerous individuals may benefit from professional mental health services, including counseling, therapy, or medication. The challenge will be to get them to agree to go to therapy and seek assistance. *Employee assistance programs* (EAPs) are imperative within the fire service. EAPs allow a third-party professional to answer calls from firefighters needing professional guidance on a range of topics. Discussions between the employee and vendor must be confidential and not shared with the fire chief or city staff. If this trust is violated, the EAP will fail.
- *Build a support network.* Surround yourself with a supportive network of family, friends, or professionals to provide a safety net. This can also help you to manage the stress and challenges of dealing with a dangerous individual. This includes getting an attorney involved to protect yourself and your department from litigation.
- *Practice self-care.* Practice self-care while dealing with dangerous individuals, including regular exercise, meditation, or therapy. You receive time off as a benefit and should use it. Trust me: you are not as well rested as you think you are.

Ultimately, the key to managing dangerous individuals is to remain vigilant, stay informed, and seek resources and support to protect yourself and those around you. Recognizing the signs of dangerous behavior and taking appropriate action can minimize the risks and potentially prevent harm. Don't delay! If it feels strange, listen to your gut. If you think you should call 9-1-1, *then do it!* Protect yourself first and act.

Chapter 8 Reflection Prompts

1. Has there been a time at work when you were afraid of someone? How was it handled? Write down what you would have done differently.
2. Is there someone at work now that you believe could be dangerous? What can you do to prevent a problem from occurring?
3. What policies and practices does your department have in place to deal with these types of people? Set a date for you and your department to discuss this issue and how to handle it.

9

Dealing with a Narcissist

Before I became a chief, I had never heard the term *narcissist*. Having lived through the challenges of dealing with several narcissists, I now recognize them easily. Individuals displaying narcissistic traits can be highly manipulative, exploitative, and dismissive of others' emotions, needs, and achievements. Narcissism is marked by an inflated sense of self-importance, need for constant admiration, and lack of empathy.[1]

Understanding how to identify and manage interactions with narcissists is essential for safeguarding your well-being and organizational health. This chapter explores the dynamics of narcissistic behavior with a real-world scenario and defines key characteristics, discusses potential causes, and offers strategies to address and cope with such people.

Scenario

You and the local mayor are at a community event to recognize three firefighters who saved a young man from drowning during a water emergency. While speaking with the mayor, you see he is scanning the room and his attention is elsewhere. In midsentence, the mayor dashes away to talk with the mayor of a neighboring community, who has just arrived.

During the presentation, the mayor spoke about how he balanced the budget so that firefighters could have the necessary equipment to save lives. He spoke about the upcoming election and the need to get out and vote. Whenever you talk to him, he does nothing but talk about himself, look down upon you, and boast about his great work. In his mind, he is the greatest thing that ever happened to your community. Before becoming mayor, he was unemployed; in addition, he had previously served as a council member and gained a reputation for deflecting blame onto city staff, rather than taking personal responsibility for his actions.

> To better understand some of the characters within our fire stations, I became an affiliate of the Workplace Bullying Institute. The founders, the psychologists Ruth and Gary Namie, have studied workplace bullying for decades and are leading authorities on the subject. During my training and research at the institute, I found this personality disorder familiar from personal experience.

Defining Narcissism

Dealing with narcissists can be challenging as they often display manipulative, arrogant, and entitled behavior. Nevertheless, correctly identifying and handling a narcissist is crucial to protect yourself from their toxic behavior.

Common characteristics of narcissistic individuals include:

- *Inflated sense of self-importance (grandiosity).* Narcissists believe themselves to be more important than everyone else. They exaggerate their accomplishments, highlight their role in every success, and expect recognition even when undeserved. In the fire service, this can show up as someone taking credit for a rescue or project that was actually a team effort.
- *Lack of empathy.* One of the most damaging traits of narcissism is the inability to recognize or care about the feelings of others. Narcissists minimize struggles, dismiss concerns, and rarely offer genuine compassion. Their focus is inward, leaving teammates feeling unseen and undervalued.
- *Constant need for admiration.* Narcissists thrive on attention, compliments, and being in the spotlight. They grow frustrated or angry when they don't receive the praise they expect. This often pushes them to dominate conversations or redirect attention from others back to themselves.
- *Manipulative behavior.* Narcissists often use charm, deflection, or deception to get what they want. They are skilled at twisting situations to their advantage, sometimes making others feel guilty or responsible for problems the narcissist created. In a team environment, this creates distrust and constant tension.
- *Sense of entitlement.* Narcissists feel entitled to praise and admiration from those around them. (For more on entitlement, see chapter 7.) This

sense of entitlement creates unrealistic expectations about how others should treat them and often leads to frustration or anger when those expectation aren't met. It's the belief that rules and standards that apply to everyone else somehow don't apply to them.

- *Fragile ego behind the mask.* Although they present themselves as confident and untouchable, many narcissists actually carry deep insecurities. Their arrogance often masks a fragile self-esteem that cannot handle criticism. Even small amounts of feedback can trigger defensiveness, blame shifting, or retaliation.

Causes

Narcissistic behavior doesn't form in a vacuum. As was also true for dangerous individuals (as discussed in chapter 8), childhood experiences, emotional trauma, social environments, and even cultural influences shape the behavior of narcissists over time. Although not everyone who displays narcissistic traits has a psychological disorder, understanding the root causes of these behaviors can enable us to respond better. Common causes include:

- *Early childhood environment.* Narcissistic traits may develop from specific parenting styles, such as excessive praise without accountability, unrealistic expectations of the child's greatness, or inconsistent affection.[2] In other cases, narcissism develops as a defense mechanism in response to neglect, emotional abuse, or feeling unseen or undervalued in childhood; for them, their ego becomes a shield.[3]
- *Low self-esteem and fragile ego.* Ironically, narcissism often masks deep-seated insecurities. Although these individuals appear arrogant and self-sufficient, their self-esteem is unstable and highly sensitive to criticism.[4] This explains why narcissists react so strongly to perceived slights or challenges to their authority. Can you think of someone who gets really defensive—or even spiteful—when you challenge their decisions? You may be witnessing narcissism.
- *Social reinforcement.* Modern culture frequently rewards narcissistic behavior, especially in leadership, politics, and media. Individuals who seek attention, dominate conversations, and self-promote are often mistaken for leaders; in fact, we put narcissists in charge many times, because it is easy to confuse charisma with character. As many of us know, humility can reinforce your leadership ability; however, a narcissist sees humility as a weakness. They operate based on more of a "me"

and less of a "you" mentality. When organizations and society reinforce narcissistic behavior, it flourishes.

- *Biological and psychological factors.* Although not fully understood, genetic predispositions and neurobiological differences may contribute to narcissistic personality traits.[5] In particular, *narcissistic personality disorder* is recognized by the *Diagnostic and Statistical Manual of Mental Disorders*[6] as a clinical mental health condition, meaning that it may also be influenced by underlying psychological and neurological development.

Addressing the Issue

The following are strategies for addressing narcissistic employees:

- *Set clear boundaries.* It is essential to establish clear boundaries with a narcissist to protect yourself from their toxic behavior. Make it clear what behavior is unacceptable and the consequences if they violate those boundaries. This can be hard to do if the narcissist has positional power over you; if that is the case, stand firm on your own morals and values. Don't lie or change your ways for narcissists. If their ways are starting to affect your health, do what is best for your mental and physical health. That may mean leaving the organization. Above all, avoid becoming a target, which can be painful.
- *Avoid engaging in drama.* Narcissists often play games to manipulate and control others. For example, they may encourage you to stand in front of them, to absorb blame that would otherwise be directed their way. Do not engage in their drama or become a pawn in such blame games. Instead, focus on your own needs and priorities. Life is about you, not them.
- *Do not enable them.* Narcissists may try to manipulate or guilt-trip you into doing tasks for them. Do not enable their behavior by giving in to their demands. Instead, encourage them to take responsibility for their actions and behaviors.
- *Seek professional help.* If dealing with a narcissist is causing you significant distress or impacting your mental and physical health, you should seek professional help. A therapist can help you to develop coping strategies and provide support when navigating difficult situations. Fortunately, receiving help is seen as a positive self-improvement

method today—so no worries! Ultimately, you are not likely to change this personality.

- *Document, document, document.* Take note of what is going on, and write your observations down in a journal. Remember, if it isn't written down, it may not have happened. Be prepared!

Identifying and handling a narcissist can be challenging, but protecting yourself from their toxic behavior is essential. With these strategies, you can effectively deal with a narcissist and protect yourself from the toxic environment they can create. They want you to believe that they are in control and better than you. They love themselves so much that they can't feel the emotions of others. In my opinion, this is a terrible trait to have working in local government, where our role is serving others. By contrast, narcissists are in the business of serving themselves, not others.

Chapter 9 Reflection Prompts

1. How can you recognize narcissistic traits in others? What are some strategies for dealing with a narcissist?
2. Have you dealt with a narcissist in the past? How did you handle the situation, and what could you have done differently?

10

Dealing with Defamation

Defamatory (false or harmful) statements damage the reputation of another person or group and destroy reputations. In the fire service, your name is your brand; it is the legacy you leave behind long after the smoke clears. Every word we say or write and every post, like, or share can either build or burn that reputation. In this digital age, where social media gives everyone a megaphone, a few hastily typed words could ignite a storm—damaging trust, ending careers, and even destroying entire organizations.

This chapter explores the growing issue of defamation in the fire service. Sometimes, the most dangerous weapon in the firehouse is not a tool; it is a tongue or a keyboard. Whether at the kitchen table or online, defamation undermines morale, fractures teams, and erodes the public trust we fight to uphold.

Firefighters pride themselves on having courage under pressure. Courage also means standing up for what is right when gossip turns malicious or when a colleague becomes the target of lies. Every person within the organization, no matter what their rank, has rights. In today's world, you don't have to look far to see people being sued in response to the incivility or harm they have caused to another's reputation and health.

Scenario

There is a lot of chatter going on at Station 1. Travis is upset that Engine 1 will be reassigned to Station 6 to improve coverage for a newly constructed 110-unit senior living facility. Camila, who served on the apparatus placement committee, supported the move, explaining that Engine 1 carries the advanced life support equipment and will likely handle most of the medical calls that occur at the facility.

During shift, Travis vents to his crew: "If the engine is going, she should go, too." After work, his frustration spills online. He posts memes mocking the department's leadership and comments that "Camila is in bed with the chief."

> Camila learns of the post through friends, screenshots the content, and submits it up her chain of command. What started as frustration over an apparatus reassignment has now turned into a serious ethical and legal problem—one that could cost a firefighter his job and the reputation of both victim and perpetrator.

Defining Defamation

Defamation is the act of making a false statement that injures another person's reputation. Two primary forms exist: *slander* (spoken defamation) and *libel* (written or digital defamation). For a statement to be considered defamation, it must meet four criteria (also see your specific state statutes):[1]

1. The statement was false
2. It was communicated to others
3. It caused harm to the person's reputation
4. It was made with negligence or malice

In the fire service, defamation can occur when someone spreads false rumors about a coworker's conduct, questions their integrity, or posts unverified information that damages another's standing within the department or community. Some firefighters confuse free speech with freedom from consequences. While everyone has the right to voice opinions, those rights end when false statements harm others or disrupt the workplace. The Supreme Court of the United States reaffirmed this in Pickering v. Board of Education, ruling that public employees' speech is protected only if it does not interfere with workplace harmony or the employer's mission.[2]

In this context, *harmony* comprises the effective and cooperative functioning of the organization. This means maintaining discipline, respect, and teamwork—all of which are essential to the public safety mission. If an online post causes the crew to stop trusting their captain or the community to doubt the department's professionalism, that speech has disrupted workplace harmony.

Think about harmony as you would about a well-trained engine company on scene. If one firefighter ignores orders or undermines their officer, then the entire operation suffers. The Supreme Court recognized that the same principle holds for public organizations: Without harmony, you can't serve effectively.

The Rise of the Keyboard Commander

The modern firehouse has a new type of loudmouth: the keyboard commander. Such people speak boldly but from behind a screen, posting memes, half-truths, and criticisms that they would never dare say face-to-face. Fueled by the illusion of anonymity and the instant validation of likes and shares, they often mistake sarcasm for humor and freedom of speech for permission to harm others.

A keyboard commander can destroy trust within a department faster than any fire can consume a building. They undermine leadership, spread disinformation, and rally others online to join the negativity. What they fail to see is that their actions live forever. Screenshots, digital footprints, and archived posts can be used as evidence in internal investigations and civil court cases.

In the fire service, courage is defined by integrity, not by how loud you type. Real firefighters confront issues face-to-face, not on social media sites or via email.

Causes

Understanding why some people defame others helps us to prevent this behavior. According to WBL's national survey: "The bully's defamation precedes the transfer. It is unconscionable that employers compel victims to suffer job loss in addition to months or years of unremitting episodes of abuse."[3] In practice, defamation is less about factual truth and more about power, emotion, ego, and organizational insecurity. The following are reasons why people may turn to their keyboard, rather than confront issues face-to-face:

- *Anger or resentment.* When people disagree with a decision, they may vent instead of addressing the difference of opinion professionally. What starts as frustration can quickly become character assassination.
- *Attention seeking and ego.* Some crave validation from peers or followers, posting shocking or harmful statements to feel powerful.
- *Lack of accountability.* In organizations where rumors are tolerated or discipline is inconsistent, defamatory behavior becomes normalized.
- *Poor leadership and communication gaps.* When leaders fail to communicate clearly, misinformation fills the void.
- *Generational and technological differences.* Some firefighters blur the line between personal and professional spaces online, unaware of how their posts affect their own credibility and that of their organization.

Addressing the Issue

Defamation must be addressed promptly when it surfaces. Just as smoke indicates a fire, harmful words signal a deeper problem within the team. Ignoring defamation allows it to spread, resulting in damage to morale and trust within your organization.[4]

Firefighters are known for confronting danger head-on, and responding to defamation should be no different. Regardless of whether defamation occurs online or inside the fire station, leaders and peers must act decisively to protect the integrity of the department and everyone in it. Through a culture that has no tolerance for defamation, the behavior can be mitigated immediately when it is identified.

The following are a few ways to address defamation:

- *Acknowledge the behavior immediately.* When you first hear or see something defamatory, whether a verbal statement or an online post, acknowledge it. Otherwise, silence yields power to the offender. If you are in a leadership position, speak with the individual(s) involved privately to verify facts and begin documentation. If you are their peer, avoid responding emotionally; instead, simply say, "That's not true, and it's not helping anyone," and then report it through the proper channels. Immediate recognition sends a message that defamation will not be tolerated.
- *Document everything.* Evidence is critical (if it isn't documented, it didn't happen). Save screenshots, text messages, and emails. Record dates and times, and compile witnesses' statements. Documentation ensures accuracy and provides protection if disciplinary action or legal proceedings occur later. Use a journal to record discussions, concerns, and actions. Without accountability, defamation becomes standard practice. When I first looked into Stan's personnel file, it was curiously thin and didn't tell the real story of what was going on. If I hadn't been told about the manila envelope covered with evidence tape in the back of the cabinet (see chapter 2), I would have had little evidence to support his termination—which would be a big problem. The importance of documentation cannot be overemphasized.
- *Follow the chain of command.* Address the issue according to your department's established communication and disciplinary procedures. The person targeted by defamation should never have to navigate the issue alone. When handled properly, the chain of command ensures fairness and keeps personal emotions from influencing decisions.[5] If

defamation involves a supervisor or officer, use the next level of command or the city's human resources department to ensure objectivity. Departments without a clear process should create one, with emphasis on confidentiality, respect, and accountability.

- *Evaluate intent and impact.* Not every negative statement is defamation, but even careless comments can cause harm. Leaders must evaluate whether a potentially defamatory statement represents an opinion, a misunderstanding, or deliberate falsehood. Intent matters, but so does impact. A firefighter's public comment might be protected as free speech, but if it disrupts trust, divides crews, or damages the department's reputation, then it violates policy and represents a departure from professional ethics.[6] The goal is not censorship; it's accountability.

- *Enforce policy and discipline fairly.* Once a given communication is determined to be defamatory, leadership must respond with consistent and fair discipline according to department policy. Selective enforcement destroys credibility; regardless of whether the offender is a new recruit or a seasoned captain, standards must apply equally. Furthermore, transparency strengthens leadership trust. Every firefighter should know that professionalism and respect are nonnegotiable expectations. Standards must apply equally.[7]

- *Educate and prevent.* Most defamation stems from frustration, misunderstanding, or lack of communication. Turn this behavior into an opportunity to educate personnel. Incorporate sessions on professional communication, ethical conduct, and social media use into regular training. Provide real-world examples (e.g., case law and outcomes from recent cases), discuss consequences of online misconduct, and teach the difference between freedom of speech and professional responsibility. Prevention through awareness is always better than discipline after the fact. As the proverb says, an ounce of prevention is worth a pound of cure.

- *Ensure psychological safety.* A healthy workplace is one where open conversation is encouraged, without fear of ridicule or retaliation. *Psychological safety* means that firefighters feel comfortable raising concerns before frustrations turn into online venting or malicious gossip. Leaders must be listening. A truly safe culture isn't without disagreement but one where disagreement is handled respectfully and internally. When firefighters feel heard, they are far less likely to harm others through words or posts.[8]

- *Model the standard.* Leaders set the tone of professionalism. If officers gossip with malicious intent, post sarcastic comments online, or speak negatively about others, the rest of the crew will follow—especially when these come from an influential teammate whom people respect. Hold yourself to the same standards you expect others to follow. When you make a mistake, own it and correct it publicly. Again, transparency builds credibility and reinforces the culture of accountability that you want to see reflected within your team.

Whenever defamation occurs, healing the team is essential. Leadership should counsel those involved and facilitate open discussion when appropriate. Reaffirm organizational values through education, and provide support for both the target and the team. In severe cases, mediation or professional counseling may be necessary.

Any time one's reputation is being attacked, it can cause pain. In severe cases, harassment, bullying, and reputational attacks can trigger PTSD symptoms. This can happen even when no physical harm has been done.[9]

Defamation is like a flashover, starting small but quickly consuming everything around it. Your experience as a leader determines whether the behavior becomes a learning opportunity or leaves a lasting scar on the department's reputation. The best defense is a strong culture built on communication, trust, and respect. When people know the truth, rumors lose their power.

Chapter 10 Reflection Prompts

1. What are the key features of defamation? How will your communication change as a leader—and as a follower—now that you possess this knowledge?
2. Have you had something said about you (verbally or online) that was not true and could harm your reputation? If you have, reflect on how the false statement(s) made you feel in terms of motivation, emotional health, and/or trust.
3. List three ways leaders and peers can mitigate defamatory statements before they spread.
4. What policies do you have within your organization to ensure that gossip does not become defamation? If there is not such a policy, create it and disseminate it to all employees.
5. How can you model professionalism and integrity—both in and out of uniform?

11

Dealing with a Workplace Bully

When people think of a bully, they may picture a child on the playground pushing others around. However, bullying extends far beyond school grounds. It appears in various forms and settings, even workplaces, where the consequences can be equally damaging. In the fire service, bullying is often overlooked or mistaken for harmless teasing or rites of passage, yet it can have severe repercussions for both individuals and the organizational culture.

After my experience with Stan, I sought to discover why I had previously been unaware of the complex personnel issues I was experiencing inside the fire station. From this research, I learned that I was not alone, as a simple online search yielded numerous results about bullying in the workplace.

Workplace bullying resembles the playground variety, just at work. We assume the workplace to be grown-up and professional, but that is not often the case. The bullies of the past become the bullies of today, as colleagues, managers, and community leaders, among others. Some never grew out of their bad habits, and over time, these bullies create havoc inside organizations.

As a new fire chief hired from outside the organization, I hoped to have time to familiarize myself with the culture and practices of the fire district but instead got caught up with a bullying situation right away. As described in previous chapters, Stan's mode of intimidation was to use his size and body language—yelling at and even poking you if close enough—to make others fear him. After this situation reached its tragic end in a murder-suicide, I started an online survey, asking firefighters if they had also experienced workplace bullying.[1] I was surprised to learn how prevalent it was in the fire service: over 5,000 firefighters from all ranks across the United States and Canada reported experiencing similar challenges with people inside the fire station.

One survey questions was as follows: "The goal of an adult bully is to gain power over another person and make himself or herself the dominant adult. They try to humiliate victims and show them who is boss. In your career, have you ever been bullied by an adult?" Over 90% of respondents indicated that they had—and for over 60%, this had happened in the past 12 months. In addition, most indicated that their workplace had done nothing about it. Sound familiar?

Scenario

Adam was recently hired as a firefighter and was excited to become part of the elite team he had seen and admired in his neighborhood. He grew up near the fire station and used to watch the firefighters washing their trucks and training outside while he walked to school. It was common for him to get a big wave from the firefighters as he walked by, and their kindness led him to want to become one of them.

After graduating from high school, Adam took firefighter and emergency medical services (EMS) classes to prepare himself for the next opening. He was ready to apply at his local fire department within a year. After the interview process, he was selected as one of three candidates to get hired. Adam was excited to begin his lifelong dream of becoming a firefighter, especially after being assigned to the busiest engine company in town.

Within his first month at the fire station, Adam felt unwelcome. Several firefighters yelled at him for not knowing where the equipment was on the engine. They teased him, asking if he fell asleep during the fire academy. The firefighters made Adam run around the truck for an hour a day and call out where each piece of equipment was. Adam was lean, so the firefighters put corn inside his turnout boots to represent his "chicken legs," implying that he was not fit enough. This lasted a week. Adam felt like a failure and began to make mistakes as the stress of the bantering got to him. Ultimately, it was clear to Adam that what he had seen from outside the fire station differed significantly from what he was experiencing on the inside.

Adam spoke with his family and friends. They told him to talk to his immediate supervisor about what the firefighters were doing to him. After meeting with his captain and telling him what was occurring, the captain told Adam that he was a probationary firefighter and should expect to be teased and harassed. "This is what we do in the fire service," said the captain. "We all have had to go through that to earn our position on the engine." Later that day, one of the firefighters approached Adam and called him a tattletale. It was clear that the captain hadn't kept Adam's concerns confidential.

Adam felt lost. He loved the job, but it was more like a reality television show than he expected. After 6 months, Adam quit.

After leaving the fire station, Adam saw that the firefighters who had harassed him were posting about him on social media. They stated he didn't have the strength or mind-set to become part of the engine company and called him a crybaby.

What issues need to be addressed immediately? I hope you find a few. Sadly, bullying the new member is a common scenario inside fire stations, even though onboarding should progress without unnecessary stress and harm. Adam was targeted by a veteran firefighter, who enlisted others to join in. Adam became the target of a mob of bullies. This put Adam in a challenging position, especially as a new employee. Bad players in the organization tarnished his lifelong memories of firefighters and his dream to become one. The captain was at fault as well. His role includes ensuring that all team members are safe—physically and mentally—and that they work together cohesively. The captain failed by supporting the veteran firefighters and the old-school culture, leading Adam to abandon his dream job.

Defining Bullying

For research on the topic of workplace bullying in the United States, the best source to go to is the Workplace Bullying Institute. The Workplace Bullying Institute offers ways to mitigate bullying through better understanding of the subject. More important, they offer insight for victims (targets) suffering psychological trauma caused by bullies in action.

Workplace bullying is "repeated, health-harming mistreatment by one or more employees of an employee: abusive conduct that takes the form of verbal abuse; or behaviors perceived as threatening, intimidating, or humiliating; work sabotage; or some combination of the above."[2] *Bullying* encompasses the entire system, while *bullies* are the perpetrators who target individuals and cause harm. Bullies work to create intimidation and fear so they can gain control over another. If you're afraid, you may do what the bully says and even give in to their demands, so they have power and control. Bullying occurs within all types of relationships, including peer-to-peer interactions. In my survey, most respondents indicated that perpetrators were their peers, rather than officers.

Bullying can come from an individual or occur as *mobbing*, which is when the perpetrators are a group of people who target an individual with repeated aggression and hostile behaviors. They may talk negatively about the target, yell at them, belittle them, or even physically attack them. Like a pack of wolves, they surround you, gradually wear you out, and take you down.

Bullying is a nonphysical form of violence but can create emotional harm and, over time, physical illness. Being a target of bullying, whether from an individual or mob, can cause severe emotional duress. In turn, this can lead to loss of productivity, physical and emotional stress and health issues, and reduced organizational cohesion.[3] Trust me: it affected my own health.

Synonyms for bullying include harassment, intimidation, coercion, aggression, victimization, persecution, oppression, abuse, torment, maltreatment or mistreatment, taunting, belittling, humiliation, and threatening. Although these words may have similar meanings, they have distinct connotations in different contexts.

Many traits of difficult people within the firehouse fit within the system of bullying. Bullying can come in many forms, including:

- *Verbal.* Yelling, swearing, sarcastic comments, teasing, and using demeaning words and phrases
- *Physical.* Pushing, poking, slapping, pulling; this can quickly lead to physical assault as well
- *Emotional.* Creating unrealistic expectations so the employee fails; constant criticism and public display of incompetence; sabotage and withholding information; gossip, rumors, and untruths, as well as practical jokes and ridicule
- *Offensive nonverbal conduct.* Body language such as thrusting, hand gestures, and facial expressions
- *Technological.* Bullying using social media platforms, such as posts, videos, tags, and pictures
- *Withholding information.* Withholding critical information so that you must seek it from them, which gives them control over you

Stan used his seniority and physical stature to manipulate others, and his bullying increased in severity over time, culminating in an unthinkable tragedy. Given our roles, he was bullying me from the bottom up. This is different than the chief bullying from the top down.

Causes

While workplace bullying occurs across many industries, the fire service presents unique cultural and hierarchical factors that can foster such behavior. These include:

- *Tradition(s).* The tough-guy mentality long cultivated in the fire service has led to acceptance of hazing or harsh initiations, excusing them as "tradition." Today, we recognize these behaviors as harassment or even assault. In many instances, those in leadership positions were themselves bullied and carry that cycle forward. In a 2021 survey, the Workplace Bullying Institute found that 61% of bullies are bosses, highlighting that authority figures are frequent perpetrators.[4] Bullying perpetuated by those in positions of power and authority reinforces the status quo through intimidation and *tradition-based tolerance.*[5]

- *Hierarchy and command.* As a paramilitary organization, the fire service generally discourages lower-ranking personnel from speaking up—for fear of disciplinary action, being labeled as not a team player, or even losing their job. When rank is misused, it becomes a tool for coercion, rather than inspiration.
- *Team dynamics.* Strong, tight-knit teams sometimes turn on those who don't fit in. This can grow into mobbing, where the entire group turns against one individual to preserve the status quo. Bullying often stems from group dynamics that reward conformity and punish difference.[6]
- *Fear of retaliation.* Firefighters work closely with one another in a family-like unit. Speaking up against a colleague can be viewed as betrayal, which discourages reporting and allows harmful behaviors to persist. In this environment, bullying thrives in silence. The culture of solidarity and fear of retaliation keep it hidden, allowing the problem to continue unchecked.[7]

Addressing the Issue

So how do you identify and curb workplace bullying in fire stations? The key is to recognize that bullying is occurring and identify the type of threat you have within the group. If you know the behavior is wrong, then you must act. Everyone on the station floor is responsible for ensuring that each of your team members is given equal respect and dignity. When you fail to stand up for those being bullied or participate in gossip and rumors, you become a perpetrator or co-conspirator, which today could cost you your reputation, job, or a lawsuit. Failure to act is an action by default and, as such, can be held against you.

You can . . . watch it happen.
You can . . . let it happen.
You can . . . make it happen.

Workplace bullying is not illegal yet in the United States; however, it is in Sweden, France, Germany, United Kingdom, Australia, New Zealand, and Canada. Even though a law prohibiting workplace bullying in the United States has not passed to date, you can still be sued for bullying, and your workplace can be held responsible if evidence proves that nothing was done to prevent it. The best way to prevent bullying in the workplace is to establish a stand-alone policy on the topic and build a culture that does not tolerate it. The policy should be built on a foundation of restorative justice (an approach that focuses on repairing harm, building trust, and restoring relationships rather than simply punishing offenders) and focused on corrective procedures. There must be no delay once workplace bullying is identified, and anyone found to have engaged in or ignored the behavior, regardless of rank, should be held accountable.

Rapid enforcement will create accountability and enhance your culture's core values and purpose. Everyone has a right to come to work and feel safe, respected, and needed.

For those who want change but know the policy has not been enforced in the past, lean in and push to add equality, respect, and positivity within the organizational culture. Develop a plan with others to attack the negativity even while acknowledging that it will not be easy. Once you get to the root cause (typically one or two people), keep pushing for positive change. Silently, everyone on the team is rooting for you as they know the attitudes need to change. If you're the chief, remember this: most of the swords coming your way are focused on your rank and ability to act. It is not about you personally; rather, it is about their insecurities, needs, wants, and so forth. Stay strong and search for those who will stand up with you for what is right.

Once you learn to recognize workplace bullying, you can take action to eliminate it within your organization. Leadership, accountability, and cultural change must be the pillars of this effort. Failure to address bullying will not only ruin the cohesion of your team but also lead to retention issues, lawsuits, and a tarnished reputation.

The following are strategies for addressing workplace bullying:

- *Set the tone.* Leadership defines the cultural expectations of the organization. If leaders tolerate bullying, it will take root throughout the culture. By contrast, if they create respectful workplace policies and enforce them, that will set a high standard and clear expectations for the workers. I implemented such a policy immediately after the murder-suicide. At each new employee orientation, I held firm on my standards, letting employees know what actions could get them in trouble on the job—in particular, harassment and treating another with disrespect. There was no question. Over time (2014–2018), that led to an increase in female firefighters joining my team (30% of the 75 firefighters in our district were women) and our being awarded the Minnesota Fire Department of the Year in 2019.

- *Take immediate action.* Taking no action is still an action—and one I see all too often. As supervisors, we want to be liked and may find it hard to deal with conflict. In this area, I found my own education lacking. What would you do if someone walked up to you and angrily called you a "terrible fucking leader"? Because they don't teach that in college, we have no mental models to draw from when needed. Bullying must be confronted in real time no matter how big or small the incident is or what color helmet the perpetrator wears. All incidents need to be documented, including the date, time, location, people involved, witnesses, and what was said and done. A verbal discussion should still be a documented one.

- *Build a culture of psychological safety.* The fire station is supposed to be a place that provides safety and comfort for all those who work there. If we truly are brothers and sisters, we should not make the station a battleground. Leaders can halt this behavior by providing training so that others can recognize bullying, encouraging everyone to participate, and stopping all past practices of hazing, intimidation, harassment, and old-school antics that no longer project a favorable view onto the fire service.

- *Support targets of bullying.* At the Workplace Bullying Institute, we learned that if you become the target of bullying, you will experience it alone. Anyone—regardless of whether a coworker or a leader—experiencing this needs immediate support. Why do we allow such behavior in one of the most trusted organizations in the world? Support your people! If you don't know what to do to help them, find someone who does and learn. Don't let your inexperience allow bullying to occur. We must act and stop the madness.

After reading this book, find ways to focus solely on the effort of keeping employees physically and psychologically safe, and your retention will improve, and the culture will flourish. As risk management expert Gorden Graham once said, "If it is predicable, it is preventable."[8]

Chapter 11 Reflection Prompts

1. Does your department still conduct initiations? Did you have to experience one to become part of the team?
2. How can you identify a bully inside your firehouse? What can you do to prevent others from becoming the target of a bully?
3. Given what you know now, could you be labeled as a bully—either now or in the past?
4. As a firefighter, have you been yelled at, sworn at, made to feel small, or gossiped about? Did you feel this was bullying?

12

Dealing with Disgruntled

Disgruntled employees are like a slow leak in a fire hose: you might not notice anything right away, until it bursts under pressure, putting the team at risk. Their negativity spreads, dragging down morale and productivity—and sometimes even compromising safety. The challenge isn't just that they're unhappy; it's that their dissatisfaction ripples through the organization, affecting everyone around them. The question becomes how to manage this behavior before it provokes a full-blown crisis.

As a leader, your job isn't to fix every problem for every person but to create an environment where people can either improve their attitude or be helped to move on. No organization thrives when it ignores negativity, because it will fester and spread like a virus, infecting good employees and creating a toxic work environment.

This chapter provides strategies for you to recognize disgruntled employees, understand their impact, and take proactive steps to help them to either improve or move out. In leadership, silence is not an option when negativity takes root.

Scenario

Havier has been with the fire department for 8 years. Several colleagues have noticed that he constantly complains. Today, he started complaining about the weight of a woman he had to help using a lift assist. He thinks lift assists are not part of the job of firefighter! He also complained about his most recent performance review and expressed the opinion that his review should not have been completed by the battalion chief, who is "never around." Over time, Havier has distanced himself from his fellow firefighters and has nothing good to say about the department, even though it has everything to offer him. Havier has become disgruntled and consistently brings down the team's mood when he is around. The sun could be brighter than ever, and this guy would find a way to make it dull somehow.

Defining Disgruntled

A disgruntled employee is an individual who is unhappy about, dissatisfied with, or resentful toward their job, employer, or work environment and expresses that discontent in ways that undermine the team. Their behavior can range from verbal complaints and resistant attitudes to more damaging actions, such as gossip, blame shifting, providing misleading information, or even sabotage. Often, their negativity surfaces daily, even during moments of success or celebration, creating a toxic environment that drags down morale and productivity. Common characteristics include resisting change and clinging to "the way we've always done it," withdrawing from team activities and showing low engagement, refusing to accept responsibility while casting blame on others, distrusting leadership and assuming bad intent, and persistently criticizing policies, leaders, or coworkers without offering solutions. Gradually, these individuals wear on the whole organization and jeopardize its forward motion. If their behavior is allowed to continue unchecked, they will plant seeds of discontent, and a garden of challenges will grow throughout the team.

Common characteristics of a disgruntled employee include:

- *Resistance to change.* Disgruntled employees often push back against new policies, practices, or tools because they prefer the status quo. They interpret change as an inconvenience, rather than an improvement, and slow progress for the entire team.
- *Withdrawal and low engagement.* Instead of participating fully, disgruntled employees often disengage, pulling back from team activities and leaving others to carry the load. Their lack of energy or enthusiasm is contagious, dragging down the morale of those around them.
- *Blame shifting and excuses.* Rather than accept responsibility for mistakes or shortcomings, disgruntled employees are quick to point fingers at others. They protect their own image by throwing teammates under the bus, creating division and distrust.
- *Distrust in leadership.* These individuals assume that leaders are incompetent, unfair, or out to get them. They spread skepticism about divisions, policies, and directions, creating an undercurrent of cynicism in the organization.
- *Persistent criticism without solutions.* Disgruntled employees are quick to complain about problems but rarely contribute ideas to solve them. Their negativity focuses on tearing down, rather than building up, leaving others frustrated and worn out.

- *Daily negativity.* Perhaps the most defining trait of a disgruntled individual is consistency. Their negative attitude shows up every day, even during celebrations or victories, casting a shadow over the team's achievements. This mind-set quickly gets old and exhausts teammates.

Causes

Disgruntled employees may have several reasons for their negative feelings at work, including dissatisfaction with their work tasks, lack of recognition or rewards, poor relationships with their colleagues or supervisors, or conflicts with company policies or values. Issues outside work can be a factor as well. It is important to look for changes in demeanor and attitude and determine if they represent a short- versus long-term mind-set.

The following are three main reasons why someone may become disgruntled:

- *Lack of recognition or perceived injustice.* Employees who feel under-valued or unfairly treated are more likely to become disgruntled.[1] When performance is overlooked or discipline feels arbitrary, resentment festers. This is particularly true in the fire service, where tradition and hierarchy can overshadow transparency and fairness. We all wish someone would have said "Good job" about a fire that was knocked out quickly or a simple "Thank you" for doing more than expected. However, rather than short-term slights, more damaging is when someone develops a long-term perception that they are being targeted.
- *Poor leadership and communication.* Ineffective leadership, unclear expectations, or inconsistent messaging can fuel dissatisfaction. When firefighters don't understand the reasoning behind decisions or when their leaders are absent or inconsistent, the gaps may get filled with negative assumptions.[2] Leadership is the key to all success within the organization—at all levels, not just the top. Fire officers are part of the leadership team and report to the fire chief, and when an officer within the ranks fails to support the leadership team, they become an immediate threat to the cohesiveness of the group. If you wanted to compound the issue, you could have a disgruntled fire officer lead a team. Furthermore, communication is an area all organizations need to consider, and with several generations currently serving within the fire service, it is crucial to consider how each of them likes to communicate and when. The most senior members may prefer face-to-face discussion

while the newest members may want a video he can watch while running on the treadmill. Today, it is best to identify the main channel where important information will be disseminated, such as email, while also offering the message in alternative ways (meetings, monthly video updates, or live broadcasts).

- *Burnout and job stress.* High-stakes, emotionally intense work like firefighting can wear down even the strongest team members. Chronic stress without proper support or rest may lead to emotional exhaustion, cynicism, and decreased professional efficacy, all signs of burnout that can turn into disgruntlement.[3] Those who experience burnout may be less productive, call in sick more frequently, and show a lack of engagement. Burnout can happen to all of us, regardless of all ranks and years of service. When these issues are identified, you should look at the daily duties and expectations of the employee. Have they increased over time? Are they realistic? And is what we have them doing more important than their long-term tenure with us? I have worked with departments that want to do it all, so they pile on prevention efforts, data entry, and clerical duties to stations that are dispatched to a call every hour. This places undue pressure on the firefighters, which will eventually burn them out and create retention issues. Making sure the response plan continues is the most important duty of the department. (Prevention is important, but it will not happen if all your people leave.)

Addressing the Issue

Leaders are responsible for addressing the concerns and needs of all employees, including those who may be disgruntled. This is a monumental task and can be tiring, but failing to address these issues can lead to decreased employee morale, increased turnover, and legal or safety concerns. I sometimes hear remarks like, "He is always that way." Still, just because an employee is disgruntled does not mean that you should ignore their concerns. In fact, lack of response can ignite resentment and dysfunction. By contrast, when employees know that you will address the issue and document what you have witnessed, progress starts. If not, you need to provide corrective action to preserve the good.

The following are strategies for addressing disgruntled employees:

- *Listen—and hear. Listening* is the act of focusing on what someone is trying to say, and *hearing* is comprehending that message for a response. Be willing to listen to the concerns and complaints of your peers and employees and take steps to address them promptly and appropriately. Over time, disgruntled people can wear on you, but you still must hear their concerns and manage each situation.

- *Provide support.* Additional training, resources, or counseling services should be provided to help employees to overcome challenges and improve their job performance. I recommend that fire departments develop an EAP. The EAP will be run by a third-party vendor that employees can call for issues they need help with. All the information shared is held confidential from the employer. Leaders and managers see only how often the employee assistance hotline is called, not by whom or for what reason.
- *Create a positive work environment.* Promote a positive work environment by recognizing and rewarding employee contributions, promoting open communication, and maintaining a culture of respect and inclusiveness. Leaders must build a foundation of policies that explain what is expected of employees to remain in good standing with the organization.
- *Address conflicts.* Be prepared to manage conflicts and disputes between employees and take steps to prevent and resolve workplace bullying, harassment, or discrimination. What foundation do you have within your department to address these issues? What authority do your frontline supervisors have to put people on administrative leave or send them home? Without authority, supervisors are just higher-paid employees.
- *Take disciplinary action.* You may need to take corrective action, up to termination in some cases, to address severe or persistent behavioral issues that may be causing harm to the organization or its employees. Remember, discipline is used to change behavior violating the rules of your workplace. If you don't have established policies, expect chaos and a challenging time enforcing the behavior.

A disgruntled employee can negatively impact workplace morale, productivity, and safety. We all should take proactive steps to address the concerns and needs of our employees, create a positive work environment, and take disciplinary action when necessary to promote a safe and productive workplace. The longer you allow this behavior to brew, the more contagious it gets. Act now!

Chapter 12 Reflection Prompts

1. What are some traits of a disgruntled employee, peer, or boss?
2. What can you do to help them to improve their attitude and productivity at work?
3. How can you support those who are trying to improve the morale of the group?

13

Dealing with Small-Minded People

The fire service thrives on innovation, adaptation, and progress. Yet, one of the greatest obstacles to change isn't lack of resources or technology; it's small-mindedness. A small-minded leader or team member can halt innovation, create unnecessary friction, and diminish morale—simply because they don't know what they don't know. The inability to embrace new ideas, whether technological advancements or changes in workplace culture, stifles progress and can make an organization obsolete.

Being small-minded isn't always intentional. Often, it stems from comfort with the familiar, fear of the unknown, or a sense of tradition. However, as the fire service evolves, we must evolve with it. The question isn't whether change will happen; it's whether we will be ready for it. This chapter explores how to identify small-mindedness in leadership and the workforce, as well as the strategies to overcome this detrimental mind-set.

Scenario

Fire Chief Williams doesn't seem to want to listen to the needs of the firefighters. Over the past 8 years, the fire department's average years of service dropped from 15 to 5. The newer generations want to use online training to supplement learning opportunities and reduce the time they need to return to the fire station. Each year, training on blood-borne pathogens is required by the Occupational Safety and Health Administration, and everyone must come back to the fire station to sit through this 4-hour class.

Lieutenant Bryan met with the chief to share a new learning program allowing the firefighters to train anytime, anywhere, using their smartphone or tablet. The program tracks the time the firefighter spends training online and shows the test scores of the quizzes taken; in addition, the training officer can be notified automatically of a pass or fail score. Because of the program's ability to show training

> competency, the firefighters would like to start using this type of training tool for regularly required noncomplex classes.
>
> Chief Williams doesn't like this idea. He thinks technology is overrated and that firefighters must be together to hear each other's ideas and concerns. They have been in the classroom for years, and he does not intend for that to change.

Defining Small-Mindedness

Small-minded individuals have limited thinking and narrow-minded views. They may be closed off to new ideas or perspectives and quick to judge or dismiss others who don't share their beliefs or opinions. Consequences of small-mindedness include lack of growth and development and limited ability to adapt to change. This mind-set can come from the top or the bottom of the organizational hierarchy.

Small-minded individuals have their own narrow worldview. Think of someone you may know who has never traveled outside their home state or seen the ocean. They may be rigid in their thinking and dismissive of opinions or ideas that do not align with their own.[1] They may be prone to making snap judgments and may not be open to exploring new ideas or approaches. Have you met a small-minded person?

Small-minded individuals can have a negative impact on the workplace, as they may resist new ideas or approaches.[2] In turn, resistance to change can lead to missed opportunities for growth and innovation, resulting in a stagnant work environment. (When we discuss generational differences, in chapter 16, you will see the need to remain open-minded about issues pertaining to technology and personal time. After all, the more you educate yourself by reading, listening to podcasts, embracing debate, and traveling, the more worldly you become.)

Common characteristics of small-minded individuals include:

- *Resistance to change.* Small-minded individuals may fight change, preferring the status quo over new ideas or approaches. In the fire service, this mind-set can prevent progress in safety, technology, and culture. You may hear, "We have always done it this way." Instead of viewing change as an opportunity for growth, they see it as a threat, taking them out of their comfort zone. I avoid statements like "We need

to change" and instead might say, "We need to do things differently."
That little tweak makes a more persuasive case.

- *Inflexible thinking.* When someone is locked in a rigid mind-set, they
  lose their ability to see possibilities outside their own viewpoint.
  Inflexible thinkers don't adapt well and struggle when circumstances
  require creativity and compromise. Such rigidity is especially dangerous
  in the fire service, where success depends on quick adjustments and the
  ability to see more than one path forward.
- *Judgmental.* Another sign of small-mindedness is being quick to judge
  others without understanding their story. These individuals dismiss
  lifestyles, perspectives, or ideas different from their own. In today's
  diverse workforce, judgmental attitudes not only damage morale but also
  make recruitment and retention more difficult. To stay fully staffed and
  effective, leaders must create a culture where open-mindedness, respect,
  and inclusion become the standard.
- *Closed off to new experiences.* Small-minded individuals avoid what's
  unfamiliar, preferring to stick to what they already know. This limits
  growth and stifles innovation. In the firehouse, this behavior could show
  up as refusing to try new training methods, avoiding collaboration with
  other departments, or rejecting technology. The result is a team that fails
  to learn, adapt, and stay prepared for the challenges ahead.

Causes

Understanding why someone becomes small-minded, rigid, dismissive, or
resistant to change can help you to address the issue with empathy and strategy,
rather than frustration. In a world of quick information and fast-paced decisions, dealing with small-minded people can be frustrating.

Here are four common causes of small-mindedness, which show how narrow
thinking often grows from experience and environment:

- *Fear of insecurity or irrelevance.* Small-mindedness often stems from
  fear—in particular, fear of becoming obsolete or losing control. In the
  fire service, where identity and value are often tied to physical skills or
  seniority, to some individuals, technological or cultural shifts can be
  terrifying. Change threatens their grip on what they know. According to
  economist John Kotter, "People often resist change not because they
  dislike the idea, but because they fear the loss of identity, status, or
  familiarity it brings."[3]

- *Cognitive rigidity.* Some individuals simply struggle to adapt to new perspectives because of deeply ingrained mental habits. Psychologists refer to this as *cognitive rigidity,* which entails a limited ability to consider alternative viewpoints or switch thinking styles. It's not that they won't change; it's that they can't see another way. Moreover, high cognitive ability doesn't guarantee open-mindedness.[4] Rational thought and flexible thinking are separate from intelligence and often underdeveloped in many workplace cultures. I have dealt with this mentality throughout my entire leadership career. Sometimes what I thought would become the simplest implementation instead turned into chaos. Those who are so rigidly stuck in the past become problems for the future.

- *Social conditioning and cultural insulation.* Some small-minded individuals were raised in environments with limited diversity, exposure, or encouragement for growth. The worldview of someone who has never traveled or explored unfamiliar perspectives will be underdeveloped. If you have never had your beliefs challenged, you start to believe them to be facts. By contrast, the most open-minded people tend to engage respectfully in debate, endeavor to see other viewpoints, and learn about other people, places, and processes.

- *Cognitive dissonance avoidance.* When people encounter new information that conflicts with their beliefs or practices, they may experience discomfort and a psychological phenomenon called *cognitive dissonance.* Rather than adjust their views, small-minded individuals may reject the new idea altogether to avoid mental discomfort; this is because it's easier than admitting the old ones might be wrong. Small-minded people tend to avoid or rationalize conflicting information, rather than change their mind-set.[5]

Addressing the Issue

To address small-mindedness in the workplace, you can take several steps:

- *Encourage open-mindedness.* Encourage open-mindedness by fostering an environment that values new ideas and perspectives and encourages employees to explore new approaches to problem-solving. If you're a leader, show that you are open-minded by soliciting new ideas, considering alternative approaches and following new technology and organizational best practices. Quit asking others to do what you failed to learn. Take the time to learn it.

- *Provide training and development opportunities.* Provide training and development opportunities for employees to broaden their knowledge and skill sets and expose them to new ways of thinking. Build a bookstore inside the fire station, share videos, and provide mentorship. Send team members to local and national conferences. Industry events such as FDIC International provide great opportunities to learn about advancements in equipment, facilities, and policy, by hearing about what other departments are doing and seeing the vender displays. Closer to home, visit other departments to see how they operate. Share officer meetings and bring in speakers to expand the minds of your team members.

- *Encourage collaboration.* Encourage teamwork, to break down silos and expose employees to different perspectives and ideas. Older and newer generations can put aside rank and seniority to hone each other's varied skill sets. Older generations might share their on-the-job experiences and lessons learned, while newer generations could impart technological tips.

- *Lead by example.* Leaders must lead by example, modeling open-mindedness and encouraging their team members to do the same. Stay informed about the fire and EMS industries. Attend conferences to learn about new equipment, technology, and practices; read blogs, watch videos, and read periodicals such as *Fire Engineering.*

Small-minded individuals tend to be resistant to change, judgmental, and inflexible in their thinking. This can harm the fire department, curtailing growth and innovation. You can address small-mindedness by promoting open-mindedness, providing training and development opportunities, fostering collaboration, and leading by example. Encourage your team to be innovative and incorporate the latest and greatest fire and EMS knowledge into your department.

Stay current on world and industry topics to become your best. Before each training session, have your team share something new to maintain a learning culture. Attend conferences and classes—you can always learn something if your mind is open. Professional development is just that: for professionals, which includes every firefighter and EMS provider I know.

Chapter 13 Reflection Prompts

1. What are some effective ways to deal with a small-minded person in your department? How have you dealt with such individuals in the past?
2. What can you do to broaden your perspective on life and the fire service? How can you broaden your department's culture? A simple change in perspective may come from visiting other fire departments and meeting with their staff to see how they operate.
3. What views within the fire and EMS community do you feel need to improve?

14

Dealing with Deviance

A deviant employee persistently violates organizational norms, policies, or ethical standards. This behavior can range from relatively minor infractions, such as consistently ignoring certain workplace rules, to serious offenses, such as harassment, theft, or sabotage.[1] In a fire department, deviant behavior can degrade morale, damage the organization's reputation, and undermine public trust.

It's essential to recognize the early signs of deviant behavior. Subtle shifts—such as sudden negativity, rule breaking, or going rogue—can be precursors to more severe issues. Respond to these signs promptly—not only to protect the integrity of the team but also to get these individuals back on track before formal disciplinary action is necessary.

Defining a Deviant Person

Deviant behavior violates organizational norms, policies, or ethical standards. Deviant employees can create problems within fire departments, detracting from productivity, morale, and reputation. We all must identify and address deviant behavior in the workplace to maintain a safe and productive work environment.

Firefighters can play a positive role by recognizing changes in their peers' attitudes on the floor and helping them to work through their issues before they become deviant and disgruntled. This is known as *peer influence* and is the preferred way to address deviance. Once the problem gets to the administration level, policies may need to be enforced, and disciplinary action may have to occur.[2]

I'll never forget when I first discovered the power of peer influence. During my early training, we had a recruit, whom I'll call "Private Johnson" here, who was so green he barely had any life skills or common sense. His mistakes piled up, resulting in the entire team constantly suffering additional drills and never-ending push-ups. The instructors tried everything they could to correct his behavior, but nothing seemed to work. We finally sat him down and were brutally honest: if he didn't shape up and start following the same standards we were all held to, we'd recommend he be dismissed or transferred to another platoon. Our message sank in; over time, his attitude improved dramatically. This underscores how critical a supportive, yet no-nonsense peer group can be in shaping an individual's mind-set. Collectively, you all get to decide what you will tolerate, before matters move up the chain of command.

Common characteristics of deviant people include:

- *Disregard for rules and policies.* A deviant person consistently ignores or bends the rules, creating an environment where standards are optional. Small violations often escalate into bigger issues when left unchecked.
- *Negative peer influence.* Rather than lifting others up, deviant employees pull them down, negatively influencing their peers to adopt shortcuts, bad habits, or negative attitudes.
- *Disruptive or harmful behavior.* These are actions that weaken trust and harm morale—from spreading gossip and rumors and turning shifts into gripe sessions to committing outright sabotage. These behaviors can be subtle yet ripple through the entire team.
- *Lack of accountability.* Deviant employees avoid responsibility for their actions, blaming others or minimizing the impact of their behavior.

- *Erosion of ethical standards.* They place personal gain or preference above organizational values, even if it puts the mission, the team, or the public at risk.

Causes

Deviance is typically not a planned event. Instead, it is created over time, growing as a result of a combination of individual, environmental, and organizational factors.[3] The following are a few key causes I've seen firsthand:

- *Perceived injustice or frustration with leadership.* Sometimes employees lose faith in their chain of command or grow resentful if they believe promotions or policy decisions are unfair. In Karen's case, she felt strongly that a female firefighter was overlooked, so she bypassed the fire chief and went right to the mayor, violating the usual protocol. Even when the leader's decision is valid (in this scenario, choosing the most qualified candidate, already specializing as an inspector), employees who perceive injustice may lash out in ways that undermine authority or organizational processes.
- *Misalignment with organizational values.* Not every firefighter comes into the job with the same background, mind-set, or respect for teamwork. When someone can't or won't adapt, friction and deviancy result. For example, Private Johnson was a recruit whose inexperience and disregard for training routines dragged down his entire unit, prompting the crew to step in. If core values like teamwork, accountability, and safety aren't shared, deviance is more likely to sprout.
- *Lack of accountability or clear policy enforcement.* If leadership doesn't respond quickly and consistently to minor infractions, some individuals push boundaries further. This can begin with bending small rules, such as tardiness or missing paperwork, and escalate to larger violations that threaten morale and safety. Progressive discipline, paired with genuine attempts to mentor and guide employees, should be used to curb these patterns early.[4]
- *Personal stress or external pressures.* Everyone has a life outside the fire station. When people struggle with financial issues, family stress, or personal mental health challenges, it can spill over into the workplace. If there's no support system or they feel they can't reach out, those frustrations might manifest as deviant behavior, disregarding policies, undermining superiors, or even sabotaging others on the job.[5]

- *Cultural or team dynamics.* In a tight-knit environment like a fire station, the unwritten rules of the team can clash with formal regulations. If cutting corners is tolerated, that encourages individuals to keep pushing the line. Conversely, a strong culture that values accountability, open communication, and mentorship reduces the appeal of deviant behavior.

Addressing the Issue

When someone is consistently deviant, it becomes everyone's problem. Based on the severity of the issue, peers can redirect attitudes and actions to protect the organization's values and prevent negative impacts from the behavior. If the problem persists, it needs to move up the chain of command, subject to a formal discipline process. Deviant behavior cannot be ignored. If it is tolerated, it spreads. Leaders and peers alike must confront it quickly, fairly, and consistently.

Here are a few steps you can take to protect your team and your culture from deviance:

- *Develop clear and concise policies and procedures.* Create, implement, review, and analyze your policies and procedures. These are the rules for those who are on the job. Be sure the members of the leadership team are on the same page, with a shared understanding with every policy, and know the steps to realign those violating the rules. Start with coaching and mentoring and move up from there.
- *Help employees to feel supported.* Withholding information, ignoring the issues, or "retiring on active duty" creates a lack of trust and transparency. Encourage an open-door policy so that people feel safe when they have a concern, idea, or even a respectful but out-of-rank challenge.
- *Follow up.* If someone has a bad day, check in with them—ideally after any incident or at a quiet time. Let them know that we are all human but the work rules and expectations are there to support a safe and cohesive workspace.
- *Act.* Failing to act is not an option, because that risks loss of respect and lack of support from the team. In fact, it sets a precedent. The more frequently you fail to hold people accountable, the harder it will become to do so in the future. Turning back is very hard to do, which is why many leaders who allow this behavior to continue, fail to ever gain back control.

Chapter 14 Reflection Prompts

1. Do you know someone who continuously violates policy and procedures? What are effective ways to deal with the situation?
2. What safety concerns can this behavior lead to in the team's operation?
3. How can peer influence be used to correct deviant behavior before it escalates to formal discipline?

15

Dealing with a Gossip

Gossip may seem like harmless chatter, but in the high-stakes environment of the fire service, its impact can be anything but trivial. This chapter explains how simple chatter, often laced with misinformation, can create organizational dysfunction and damage reputations. This chapter examines why people gossip, how it spreads, and the steps leaders can take to address it, and it shows that preventing gossip is critical to preserve the mutual respect and clarity every station needs.

Scenario

Firefighter Storm oversees the fire academy's new-hire program, working directly with every new firefighter in training. Because he interacts closely with recruits, he learns personal details that the interview panel might never know. During a visit to Station 3, Storm told the on-duty shift that one of the female candidates is a "party girl," speculating about her personal life, and mentioned that the entire class is lagging behind on skill tests.

This may sound like harmless small talk, but it's actually gossip, which can turn your station into a breeding ground for cliques, rumors, and low morale. This situation is particularly unfair to the female candidate, who is being labeled in ways that could damage her reputation before she has even settled into the department. As a leader or peer, the key is to recognize gossip when it arises and address it quickly, so that it doesn't unravel your team's trust in one another.

Defining Gossip

People gossip for various reasons, including boredom, desire for attention or social status, or lack of communication from management. Common signs of a gossiping employee include:

- *Sharing rumors or private information.* Gossiping employees may share rumors or confidential information about their colleagues, supervisors, or the company, which can create distrust and erode morale. Not all the information you receive is intended to be repeated. Before you repeat personal information, make sure you have permission. When someone shares something personal with me, I often ask if I can share it or if they want it to remain private. Certain information, however, is exempt; you may have to disclose information if a violation of policy or the law has occurred or if there is a threat to anyone's safety.
- *Creating alliances (wolf packs).* Gossiping employees may form group alliances—cliques or partnerships—that share information or rumors, which can create division and conflict within the workplace. Like a pack of wolves, these groups can be incredibly destructive, causing long-term harm to coworkers' reputations and morale.
- *Spending excessive time socializing.* Gossiping employees may spend excessive time socializing or chatting with colleagues, which can decrease productivity and harm team dynamics. Remove yourself from the frenzy of the wolf pack so that you don't become associated with it. In other words, when you find yourself talking about others, it is probably time to go sweep the floor.
- *Engaging in negative talk.* Gossiping employees may engage in negative conversations about colleagues or management, creating a culture of negativity and cynicism. A single person can set off a chain reaction of negativity by sharing the wrong information. Gossip can negatively affect employees and the organization, through decreased morale, loss of productivity, and even legal issues. In addition, gossiping can harm the organization's reputation and create a negative image in the eyes of customers, stakeholders, and the public. Bad information with your department patch on it is doubly bad.
- These impacts are consistent with research showing that gossip and rumors act as informal communication networks that can undermine trust, distort facts, and harm organizational performance.[1]

Causes

To manage gossip effectively, you will benefit from understanding why it exists in the first place. Gossip is not always malicious; it often grows from emotional or organizational voids that leaders can identify and address. Here are four common causes of gossip especially relevant to the fire service:

- *Lack of information (communication gaps).* When leaders don't communicate clearly, frequently, or transparently, people fill in the blanks themselves. Gossip becomes a substitute for official communication. In the absence of clarity, people create their own stories, and these are usually worse than the truth. This is also known as *confabulation*, a psychological phenomenon where a person unintentionally fills gaps in memory with fabricated or distorted information that they then believe to be true. Note that this is not the same as lying; confabulators genuinely believe their version of events, even when those details are inaccurate.[2] In the firehouse, confabulation might sound like either of these statements: "I'm pretty sure the new hire failed the academy"; or "I heard the chief say . . ." Even if no such event ever happened, our brains prefer a complete story over an incomplete one; hence, when details are missing, especially in high-stress or informal situations, our minds subconsciously fill in the blanks. Over time, confabulations may get retold as if they were facts, feeding gossip and spreading misinformation.
- *Need for belonging and social connection.* Fire stations are tight-knit environments. Gossip can feel like a way to bond or gain acceptance, especially among newer members seeking connection and approval. Sometimes gossip isn't about a story; rather, it's about being part of the circle telling it. Gossip plays a role in social bonding in group settings.[3] Although natural, this becomes harmful when it is used to exclude or demean.
- *Low morale or disengagement.* When firefighters feel unmotivated, underappreciated, or bored, gossip provides distraction and a heightened sense of importance. When purpose fades, petty talk takes its place. Gossip becomes a way to vent or rebel against the organization. One study on negative workplace behaviors showed a correlation between low job satisfaction and gossiping and spreading rumors.[4]
- *Unaddressed conflict or competition.* Gossip can become a weapon when unresolved tension exists between people or when individuals are competing for recognition, promotion, or favor.

Addressing the Issue

Gossip doesn't disappear on its own. Leaders must call it out, redirect the conversation, and reinforce a culture of respect.

Here are several steps you can take to address workplace gossip:

- *Educate firefighters.* Emphasize that gossip erodes morale, damages trust, and imperils the overall safety culture. As with deviant behavior, gossip is counterproductive to the mutual respect we rely on. If you hear gossip, put a stop to it. If everyone were to stop participating in it, there will be no one to gossip to.
- *Encourage open communication.* When rumors start circulating, clarify the facts as soon as possible. Sometimes the simple act of explaining a policy or promotion decision can snuff out gossip. Provide accurate information directly from the source, ideally in person, especially if it's critical or sensitive news.
- *Discourage clique formation.* Rotate assignments, organize cross-training, and foster an environment where recruits learn from a variety of experienced firefighters. This keeps smaller factions from becoming echo chambers for rumors.
- *Address rumors swiftly.* Nip confabulation in the bud. If you catch wind that a story is morphing or ballooning, track down the original source and clarify the facts. It's better to handle it head-on than to let misinformation spread unchecked.
- *Take disciplinary action when necessary.* In severe cases, gossip can be as destructive as any other policy violation, so formal discipline, suspension, or even termination may be necessary in response.[5] This is especially true if the gossip fosters a hostile work environment or seriously damages another member's reputation.

Although not all gossip is harmful, when it targets others or when private information is shared with destructive intent and/or consequences, gossiping can create a hostile work environment and negatively affects employees and the organization. We should all take proactive steps to address the issue of gossip by educating employees, promoting open communication, discouraging cliques, addressing rumors, and taking disciplinary action when necessary. The more detailed and transparent you can be when sharing the information, the better. Clear communication reduces uncertainty, which is often the fuel that gossip feeds on. When people understand the *why* behind a decision and know what is expected of them, there's less room for speculation or rumor.

Likewise, when leaders keep teams focused on meaningful work, specific tasks, goals, and follow-up, idle time and informal chatter lose their power. Productive teams don't have time to gossip because they're too engaged in achieving results and supporting each other.

Chapter 15 Reflection Prompts

1. How can you identify a gossip in your workplace?
2. What can you do to redirect the discussion back to what is important?
3. How can you prevent gossip from occurring?
4. What are some strategies to you can use to ensure details about an incident are accurate and based on facts, not assumptions or rumors?
5. How can you show your team that gossip has no place in a firehouse?

16
Generational Differences

Leadership today is more complex than ever, not because the job itself has changed, but because the people doing the job have. Many departments have four—and some five—generations working alongside each other, each bringing their own perspectives, communication styles, and values. What worked for one generation doesn't necessarily work for another, and leaders who fail to recognize these differences risk losing engagement, productivity, and their best people.

Understanding generational differences is not about favoritism; it is about effectiveness. The way firefighters communicate, learn, and engage with their work is influenced by their experiences growing up. If we refuse to adapt, we create unnecessary barriers that lead to frustration from all sides. Although we should not put generations in silos or base our leadership style on any one generation, we can use the formative experiences of all our members to widen our view of the world.

This chapter explores how generational gaps influence leadership, communication, and team cohesion. The goal is not to divide but to bridge, to build a fire service that thrives through collaboration and respect.

work. Tim replies that he tried to call but since there was no answer, he sent a text. When Marcus notes that he did not receive a voicemail, Tim confirms that he did not leave one, asserting that Marcus should have seen that he called and called him back.

My Experience

At one point in my career as a fire chief, I had four generations inside the fire station. Initially, I did not fully understand how much that would affect me—until I started getting pushback regarding directives that seemed simple to me. For example, I remember announcing that email would be the official means to communicate fire department information. Some of those older than me had a problem with that. They would rather I type a memo and post it on the corkboard, then sign off when they read it. In fact, several stated, "I don't have email." By contrast, members in youngers generations would prefer that I make a video of myself and post it online to watch on their social media feed. Acknowledging that generational differences exist can help you to better relate to people.

Defining Generations

A *generation* is determined by the years in which a group of people are born. Over a span of 15 to 25 years, people grow up and become adults, then may have their own children, and this length of time roughly corresponds to each generation.

We will always be part of the generation we grew up in. Collectively, each generation experiences societal changes, and these form the basis for the beliefs and biases that they carry throughout their lives. When communicating with different generations, consider how their varied experiences have shaped their worldview, including how they view leadership, management, and politics, as well as the way they communicate.

Generations have names that distinguish them from each other. These commonly refer to cultural and historical touchpoints. Read on and see table 16–1 for more on each of the generations and their unique experiences and perspectives.

TABLE 16–1. Generational differences (part 1 of 2)

Characteristics	Silent Generation	Baby Boomers	Generation X
Birth years	1928–1945	1946–1964	1965–1980
Proportion of global population	3%	15%	18%
Experiences that have significantly shaped values, attitudes, and perspectives	Cold War; nuclear threat; civil rights movement; Korean War; television; rock and roll; space race; established gender roles and expectations	Economic boom; radio news; civil rights movement and Martin Luther King, Jr.; Vietnam War; hippie movement and Woodstock; assassinations; technological innovations; shift in gender roles and beginning of women's liberation	Computers; MTV; Internet; VCRs; CDs; energy crisis; globalization; recession; dual-income households; increased divorce rates; political shifts; Space Shuttle disaster; AIDS epidemic; latchkey kids
Attitude toward career	Loyalty forever; job forever	Employers define careers; title is everything	Portfolio career; loyal to profession but not necessarily to employer
Attitude toward technology	Disengaged	Early adopters	Digital immigrants
Aspiration	Own a home	Job security	Work/life balance
Communication media	Formal letter/mail	Telephone	Text message or email
Communication style	Face-to-face	Meetings and telephone	Text or email
Technological advancement(s)	Commercial air travel	Television	Home computers

Abbreviations: AIDS = acquired immunodeficiency syndrome; AR = augmented reality; CD = compact disc; COVID-19 = coronavirus disease 2019; DEI = diversity, equity, and inclusion; MTV = Music Television; VCR = videocassette recorder; VR = virtual reality.

The Silent Generation (1928–1945)

Members of this group are 81 to 98 years old. In many fire departments across the United States and Canada, it is not uncommon to find someone of this age still serving their community. Several noteworthy events that affected this group and their parents guided their cultural views.

The Great Depression (1929–1939) resulted in widespread poverty and hardship for the parents of members of this group, who lost their savings. Their

TABLE 16–1. Generational differences (part 2 of 2)

Characteristics	Generation Y	Generation Z	Generation Alpha
Birth years	1981–1995	1996–2011	2012–2024
Proportion of global population	22%	32%	Still growing
Experiences that have significantly shaped values, attitudes, and perspectives	Rise of the Internet; social media; digital radio; 9/11 and war on terror; school shootings; the Great Recession; smartphones; interconnected world; higher education and online learning; COVID-19 pandemic; mass shootings	Digital natives; social platforms; COVID-19 pandemic; climate change; postrecession; increased politics; mental health; DEI; entrepreneurial; world media access	Digital natives; constant connection; personalized learning; climate change; mass shootings; global health; family dynamics; activism; mental health
Attitude toward career	Digital entrepreneurs; work «with,» not «for»	Will move seamlessly between jobs; start their own businesses	Unknown
Attitude toward technology	Digital testers	Digital natives	Digital dependent
Aspiration	Freedom and flexibility	Security and stability	Unknown
Communication media	Text or social media	Integrated devices	Unknown
Communication style	Online and mobile	FaceTime	Unknown
Technological advancement(s)	Smartphone	AR/VR; autonomous vehicles	Tech, tech, tech!

Abbreviations: AIDS = acquired immunodeficiency syndrome; AR = augmented reality; CD = compact disc; COVID-19 = coronavirus disease 2019; DEI = diversity, equity, and inclusion; MTV = Music Television; VCR = videocassette recorder; VR = virtual reality.

parents dedicated their time to survival and rebuilding their lives. Their parents also lived through World War II (1939–1945) as either soldiers or supporters of the war effort.

A postwar economic boom sparked optimism and a sense of duty. In general, members of the Silent Generation are hardworking, loyal, and have a keen sense of responsibility. Younger members of this group grew up with the civil rights movement, which fought against segregation and discrimination that led to many of the fundamental policies in the United States today. As leaders, they commonly put forth a military-style demeanor. This generation gets the job done by use of hand tools and pen and paper. The technological advancements in their formative years were television, commercial air travel, and the development and distribution of penicillin.

Baby boomers (1946–1964)

Members of this group (also called *boomers*) are 62 to 80 years old. Like the Silent Generation, they, too, experienced several noteworthy events that shaped their worldview, including the civil rights movement, antiwar protests condemning the Vietnam War, and the dawn of the women's liberation movement. Many boomers served in Vietnam, and the war's negative impact on returning soldiers is etched in their minds. The hippie movement responded to this time in American history with a message of peace and love. During the 1960s, older members of this group experimented with drugs, such as heroin, acid, and mushrooms. They also participated in alternative lifestyles and showed how music and artistic expression could bring people together during events such as Woodstock.

Although the baby boomers experienced economic recessions, financial instability did not affect them as much as it did the generations prior. Technological advancements during their formative years included the personal computer and color television. The baby boomers are optimistic and unafraid to challenge the status quo and authorities. They had ideas and were not afraid of capitalizing on them.

Generation X (1965–1980)

Members of this group are 46 to 61 years old. This generation experienced financial insecurity, with high inflation and unemployment rates. Consequently, both parents often needed to work, leaving the kids to watch themselves; hence, they are also known as the *latchkey generation*. Although this led to a sense of independence, divorces and single-parent homes were also common.

This group experienced global terrorism firsthand, including the World Trade Center attack on September 11, 2001. Local terrorism became more frequent as they witnessed mass shootings in schools and shopping malls across the United States. Generation X saw a culture shift, with acceptance of lesbian, gay, bisexual, transgender, and queer rights after high-profile actors and singers shared their true identities, such as Freddie Mercury (of Queen) and Elton John.

Technology expanded in this generation as the Internet and home computers became more common. Email was becoming more frequently used with the advent of Yahoo! and America Online accounts. The cellular industry—and early texting devices included the BlackBerry cellphone—changed the way people communicated with one another.

This group experienced controversy over the war on terror and the masterminds of terror schemes. This eroded their trust in authority and institutions. Consequently, they may not be as loyal to organizations as the generations before them. Nevertheless, they are independent, adaptable, and creative.

Generation Y (1981–1996)

Members of this group, also known as *millennials*, are 30 to 45 years old. Many have suggested that they don't want to work. Nevertheless, this group's members now run leading corporations worldwide and are doing quite well. In fact, their experiences growing up have shaped many of the discussions about work/life balance that we have today. My older daughter is in Generation Y, and I have learned a lot through her experience.

This group grew up in a world of technology. They had cellular phones in their teens, growing up with the burgeoning Internet and using applications such as Snapchat and Facebook to communicate and get their information. A sense of fear, from terrorism to personal debt, also characterized their formative years. They witnessed school shootings and 9/11, as well as the school shooting at Columbine High School in Colorado and bombings of Trans World Airlines Flight 840, Pan Am Flight 103, and the North Tower of the World Trade Center. The cost of higher education was also exorbitant, and finding stable jobs was hard. This group participated in global climate change awareness, bringing to the forefront a discussion many thought we should have had decades ago. Now, as local leaders, they fight to save our planet.

For this generation, working is a means to an end. Unlike previous generations who have lived to work, this group works to live. Consequently, you may have trouble finding firefighters in this generation who will work overtime. I asked my grown daughter why this might be, and she replied that life is short and that spending her precious time at work is not something she wants to do. She wants to have time alone, hang out with her husband, and have tea with friends. After graduating from New York University in Manhattan, she became a nurse and worked through the pandemic in an emergency room. From the calls I received from her during that time, I can understand her need for rest and relaxation. As a child, she also witnessed her parents both working, trying to balance work with their children's activities and time together.

As pioneers of technology, globalization is in the palm of Generation Y. They are exposed to everyone's views and ideas from around the world with every touch to their screens—potentially hundreds of times per day.[1] To previous generations, their communities were limited to their immediate area and the people they surrounded themselves with; this group has a more global viewpoint, with a strong sense of community across the world. Consequently, they desire meaningful experiences, stand up for new ideas, demand change from society.

Generation Z (1997–2012)

Members of this group, also known as *iGen*, are 14 to 29 years old. This group will be the focus of our recruiting for the next decade.

Generation Z are the original *digital natives*, having been surrounded by technology their whole lives. In addition to having been issued a cellphone at an even earlier age than the millennials were, many were issued tablet computers from kindergarten on, to complete their in-class assignments and homework.

The rotary phone is unknown to Generation Z, and manual car windows are long gone. They order fast food from their phones and have it delivered by a driver. Everything is powered by batteries and needs to be charged. If you need a new pair of shoes, you can use an app, and they will arrive in a matter of hours, perhaps even be delivered by drone.

They have witnessed political instability and global conflict. Terrorism and mass shootings fill social media feeds, and gun control continues to be debated. In addition, this group has greater awareness of issues related to diversity (e.g., sex and gender, race and ethnicity). We must rise to the challenge of considering the struggles and differences of others when working with them.

They are willing to travel the world, yet the map inside the fire station is useless, as most in this generation do not even know how to read it. If their smartphone dies when the alarm sounds, they may have difficulty getting to the incident location. (This is in stark contrast with the days when we used to have classes on reading a map.)

My younger daughter belongs to Generation Z. She is a social butterfly, popular, and intelligent. She went to college to become a social media marketer and graduated from my living room as a result of the pandemic. She worked as a firefighter recruitment coordinator to initiate a statewide recruitment program for a local nonprofit representing firefighters. Then she enrolled in a county fire academy as a nonaffiliated firefighter to understand better how the fire service operates. After graduating with her certifications, she joined our local volunteer fire department and fell in love with it—so much so that she applied for a career position and was hired immediately. Watching her in the role of a firefighter has been interesting, as this generation needs flexibility. Unlike my own feeling that the world would stop if I misplaced my pager, hers lies on the table while she relies instead on a third-party app on her smartphone to notify her of incidents. The fire department is her job, and she loves the purpose it brings to herself and others. In addition, though, she runs a business after work and finds time to spend with friends, care for her dogs, and travel. So despite being her passion, the fire service is not her entire life or sole identity.

Look around your fire station. If you are using corkboard, paper, and pen or pencil, this will create difficulty for this generation. Why write it down when Siri can do it for you? I have seen how stressful the older ways of doing business are to Generation Z firefighters coming into the fire station.

Generation Alpha (2013–2029)

So, who comes next? The traits, views, and experiences of Generation Alpha—whose parents are largely millennials—are forming now, so a greater understanding of this group will come with time. Their world is full of technology and social media platforms. They survived a lockdown during the pandemic and experienced political madness that has raised questions about truth and democracy. Artificial intelligence has become a technological threat as it is used to clone voices and faces and fabricate news. Stocks and bonds are falling, and global debt has skyrocketed.

Can you predict the experiences that will shape their outlook on life and the world? I suspect this group will be diverse in their thinking, open to new ideas and change and conscious of environmental issues. However, their values may be very different from mine. Generation Alpha will be mobile and may not be able to develop a sense of community when the world itself appears to be so divided.

Integrating Generations in the Fire Department

A balance of mentorship exists in the fire service today. I remember starting at the station at age 19 years. At the time, Jack Schmidt had nearly 40 years on the job. We used to sit around the station and listen to his stories. He would mentor us from his experiences and lessons learned from previous incidents. I did not have much to offer because I was new and worked within the system and practices in place. Everything was written on paper, with a pen, while holding a clipboard.

Fast-forward to today. The initial information is populated into a records management system when a call is dispatched. Our vehicles can be tracked by automatic vehicle location, and our commands can be input using a voice-to-text application. We can drop and drag commands and symbols on a flat-screen display, automate response plans, and instantly post (live stream) the incident on social media. When we return to the station, a Wi-Fi network using near-field communication can scan your equipment inventory and print a sheet verifying that everything is in its correct place.

If none of that made sense to you, consider Generation Z and those who will follow. On day one, they can quickly enter a report into your records management system, figure out your network, set up your projector, and manage most of your department from their phone. Today, there is a balance in mentorship, and no longer is age or seniority the sole determining factor. The newest member can teach the older member how to use technology, understand their slang, and find solutions through the use of apps and automation.

Currently, we live in a world of high debt, and the thought of never owning a home is prominent in the minds of those belonging to the younger generations. Inflation and interest rates continue to be major concerns.

Mental health has become a mainstream topic in the fire service. It is also a top priority for younger generations. Financial and political insecurity have fueled anxiety, depression, and suicidal thoughts in our youth.[2] Together, we all need to support one another through the challenging times in life. Now that you understand the challenges and mind-sets of the various generations, use this information to understand the people you work with, lead, and manage.

Addressing Differences through Communication

One fundamental difference in the workforce is how we communicate. The leaders I came up under wanted to have sit-down meetings to talk things out. By contrast, younger generations communicate using technology.[3] The supervisor may want a phone call, but Generation Z wants to send a text when they are late for work. Have you ever called your kids only to get a text back, asking "What's up?"?

Here are a few strategies for communicating across different generations:

- *Adjust communication styles.* For baby boomers and Generation X, phone calls and in-person meetings are valued. Millennials and Generation Z prefer texts, emails, and instant messaging. A mix of both ensures clarity, expecting one group to fully adopt the other's preferences won't work. We all must adapt!

- *Modernize training approaches.* Training is another area that creates contention inside the fire stations. Older generations like formal classroom-style learning with hands-on training, whereas younger generations might prefer to knock a few classes out while on the treadmill using their mobile device. Similarly, older generations tend to value traditional classroom learning, while younger generations thrive with on-demand, digital-based training. Therefore, offering *blended* learning—a mix of in-person training with online options—ensures that all generations can learn effectively.
- *Find strengths instead of fighting differences.* Leveraging each other's strengths is the key to success in a diverse workplace. Being open to new lessons and experiences helps each firefighter to learn in the way that best fits them. Share on-scene scenarios with those who have not had the same experiences. What worked and what could you have done differently? Younger firefighters can share their technological skills to improve efficiencies in report entry, truck checks, and inspections, among other areas. Put companies in teams that can learn from one another. Do this on shift and during training.
- *Set clear expectations.* Let policy determine how you communicate vital information among your team. Note that electronic communication does not impart emotion and can be interpreted in a way other than how it was not intended. Consider sharing information in multiple ways: in person, through email, and via texting. Although a formal way to receive communication should be specified in the policy, using multiple ways adds redundancy to ensure that messages are received by all.

Closing Thoughts

Generational differences do not have to create division; they can create strength when understood and leveraged correctly. When leaders remain adaptable and open-minded, they can create an environment where every generation thrives. The key is not forcing one side to conform but finding common ground where everyone contributes their best.

The fire service has always been about teamwork. That does not change just because of the way we communicate or understand the world. Our work has evolved. Adapt, learn, and lead because the future of the fire service depends on it.

Chapter 16 Reflection Prompts

1. What are some ways to create an environment where people of all generations feel included?
2. How can you use generational cues to better communicate and understand one another?
3. What are you doing to make younger generations feel comfortable inside your organization?
4. What has your organization done to incorporate technological advancements that younger generations are familiar with?

17

Understanding Body Language

As a rookie firefighter, I enjoyed staying on the fire scene to assist the fire investigators in determining where and how the fire began. Theirs is a fascinating job that requires a clear understanding of fire dynamics, mechanical aptitude, technical experience, and interview skills. I enjoyed this so much that I decided to get a degree in fire science, including origin and cause investigations. One of my favorite parts of these classes was learning to read body language.

When an incident requires investigation, you use witness statements, evidence, and the scientific method to determine a hypothesis about what occurred. In many cases, the incident disposition will remain open without the testimony of witnesses and those involved. For interviews, it is vital to have a plan and to gain the trust of those you will talk to about the incident. Find an area free from distractions that is quiet and comfortable for the people you will speak to. I typically started my conversations by introducing myself and asking interviewees basic questions, such as who they are and their relation to the incident. Whenever I asked questions, I would look the interviewee in the eyes, listen (to their words and tone), and watch their body language. By focusing on nonverbal cues, you will be able to identify the emotions of the person you are talking to—and if they are similarly attentive, they will be able to read yours as well.

Since birth, we learn to understand one another without words—through gestures, facial expressions, and movement. For example, crossing your arms and crunching your eyebrows together shows someone you are mad, and crunching your eyebrows and tilting your head may signal you are trying to understand. When listeners interpret a speaker's feelings or attitudes—particularly when verbal and nonverbal cues conflict—research suggests that 55% of the perceived message comes from body language, 38% from tone of voice, and only 7% from the actual words.[1,2]

Body language also plays a vital role in managing people, as it can communicate a wide range of emotions and attitudes. As a manager or a leader, how you use your body language can impact how your employees perceive you and respond to your leadership style. With Drill Sergeant Glasgow, I could tell what he meant loud and clear just by the look on his face or his gestures.

Features of Body Language

Maintaining good eye contact is critical when managing people. Eye contact shows that you are engaged and interested in what others are saying. It also conveys confidence and sincerity, which can build trust and rapport with employees. However, too much eye contact can come across as aggressive or intimidating, so it's essential to strike a balance. Have you ever spoken with someone who appears to be looking straight through you?

Another crucial aspect of body language is posture. Good posture can convey confidence, authority, and professionalism; by contrast, slouching or hunching may make you appear lazy or uninterested. When speaking with someone, maintain an upright posture with your shoulders back and your head high; this will make you look more authoritative and help you to feel confident and in control. As an example to avoid, consider the posture of a boxer; when they fight, they place one foot forward and one in back for balance. However, this places the front of your body away from the person you speak to, giving the subconscious message that you are closed off. Instead, face the person and open your arms. This sends the message that there are no threats and you are engaged.

As simple as it sounds, give your full attention to whoever you are talking to. At a one meeting, I was speaking with the sheriff of a large county and thought we were in a deep discussion—until he saw a senator walk into the room and walked away in the middle of my sentence without even excusing himself. Not only was that rude, but it also showed me that he did not even care about the topic of our conversation; I never talked to him in-depth again. Moreover, the rise of the smartphone, somewhat ironically, has been a communication killer. Set your phone so it does not ring while you communicate. If you forget, do not pick it up if it does ring; doing so will disrupt your conversation's mood and indicates to the person you are talking to that they are not as important.

Gestures are also expressive. Using open hand gestures with uncrossed arms, palms visible, and facing forwards conveys warmth, friendliness, and enthusiasm. On the other hand, closed gestures like crossed arms can communicate defensiveness, distrust, or even hostility. Hence, use natural, open gestures that match the tone and content of your message.

When you talk to groups, video-record yourself. If your hands and arms are consistently moving, they can distract your viewers, and they may miss the meaning of your message. After teaching at one conference, I went through my class reviews, one of which said, "You touch your nose a lot." When I reviewed the video, this was true. I had been doing it subconsciously—for some reason. Now, every time I make a presentation, I think about where my hands are, the tone of voice I use, and the eye contact I make. Being an effective communicator requires careful planning and continued training.

Facial expressions are perhaps the most potent aspect of body language. A smile, for example, can instantly put people at ease, communicating warmth and friendliness. Conversely, a frown or a scowl can express displeasure, frustration, or anger. When interacting with others, be aware of the messages your facial expressions are sending and adjust them accordingly.

Finally, cultural factors and differences between sexes and generations can affect body language. Various cultures may interpret body language differently, so it's important to be sensitive to these differences when communicating with diverse audiences. Get to know your team and learn more about one another's communication styles and cultures. This holds true for both sexes and across generations. As emphasized in this book, your way of doing things will likely be different than how others work. Once you acknowledge that and find the best way to communicate with one another, you will be on the path to success.

When interpreting body language, context matters. The same gesture can have different meanings, depending on tone, timing, and setting. The photographs in figures 17–1 through 17–6 showcase different examples of body language and explain their meanings. The key is to read the full picture, not just isolated actions. If you are not sure what a speaker or listener's body language means, then ask for clarification: "Are you mad at me?" or "Are you giving me a compliment?"

Communication Assimilation

We adapt our communication style to the audiences we communicate with. This is known as *communication assimilation.* This involves adjustments to your tone of voice, word choice, and overall communication style to fit into the group you are interacting with.

In group dynamics, it is common to assimilate aspects of body language— for example, yawning when someone else yawns. If you find yourself in a serious conversation or a heated argument, remain conscious of both your own body language and that of the person(s) in front of you. Importantly, if they have their arms crossed, sending you a signal of being closed off, be sure you do not mimic their actions. As an experiment, the next time you are in a group, cross your arms when you talk and then look around to see whether others cross their arms, too. Next, open your arms and watch others do the same.

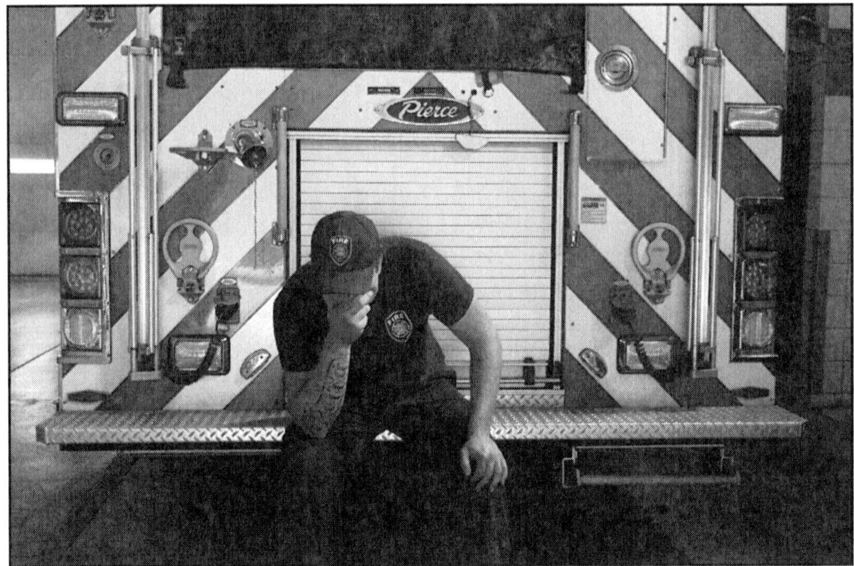

FIGURE 17–1. Head in hand: a classic sign of emotional overload. The firefighter is seated, head down, a hand covering his face. This is a strong signal of emotional strain, suggesting feelings of defeat or being overwhelmed. Approach with empathy and ask how you can support them.

FIGURE 17–2. Hands on head: processing pressure. The firefighter has her hands placed on her head, suggesting mental overload, stress, or panic. Furrowed brows and fixed expressions reinforce the presence of internal struggle. In difficult situations, recognize this as a signal to slow down, maybe even take a break, before pressing forward.

FIGURE 17–3. Fist bump: a nonverbal indicator of mutual respect. The smiling face and outstretched fist signify rapport and positive engagement. This is the body language of connection—a good sign you're on the same page emotionally.

FIGURE 17–4. Looking down, tuning out: the body language of disconnection. The firefighter on the left is absorbed with his phone and has his body turned slightly away from the speaker, signaling clear disengagement, a nonverbal cue that he's uninterested or emotionally withdrawn from the conversation. Meanwhile, the firefighter speaking at right remains engaged, using hand gestures to communicate, but most likely feels ignored or dismissed. In difficult conversations, this kind of body language shuts down trust and inflames tension. By contrast, active listening, eye contact, nodding, and facing the speaker could shift the tone and foster mutual respect. Whenever you communicate with someone face-to-face, especially about a serious matter, silence or shut off your cellphone and engage with them.

FIGURE 17–5. Open arms and smiles: defusing tension with humor and openness. Both firefighters are smiling, and the one at right is using open-handed gestures, showing transparency and good intent. An open posture invites collaboration, even in difficult discussions.

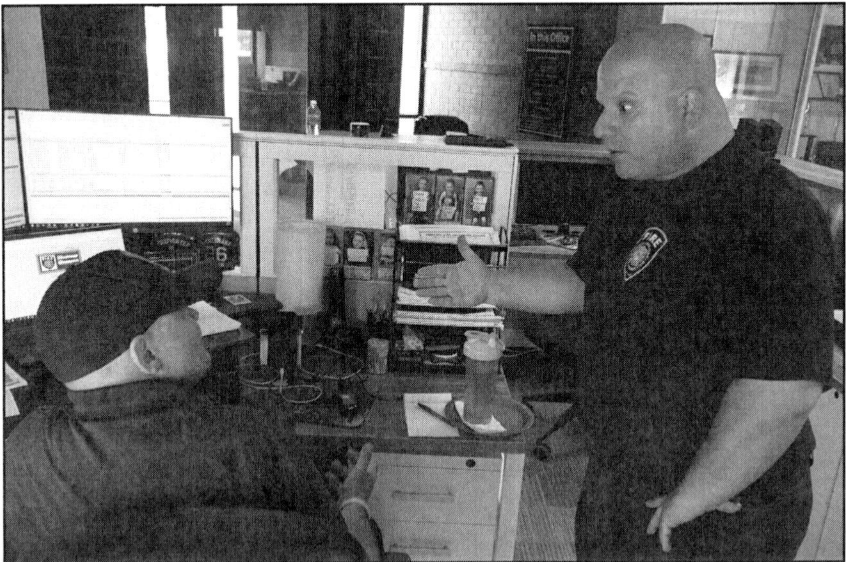

FIGURE 17–6. Open hand versus closed postures. The standing firefighter is using an open hand gesture, a sign of willingness to engage, but his intense expression and direct body angle may come across as confrontational. The seated firefighter, leaning back, appears defensive or disengaged. Their body language indicates misaligned communication styles.

Closing Thoughts

Adjust your presentation style to your audience. When speaking at elementary schools during Fire Prevention Week, I use simple words and an energetic tone of voice to add fun to my talk, kneeling to their level to minimize the towering effect an adult can have. They are in tune with facial expressions, so be cognizant of your expressions. Getting to their level and smiling makes them feel seen and safe. By contrast, in high schools, I use a more serious tone of voice and became very descriptive in my message. They understand the world more than elementary children, so it is okay to speak candidly. Look sharp to gain respect and think approachability over authority. As a further example, the U.S. Army's communication style was firm and direct but used words that may not fit well in a professional setting.

Overall, how you assimilate and use your body language can help you to communicate better. Be conscious of this when you are dealing with people who may be upset or difficult. Body language can add value to the conversation and even calm it down.

Chapter 17 Reflection Questions

1. Watch yourself tell a story in the mirror. How do you use your eyes, eyebrows, and body when telling a story? Are you using filler language, such as "um," "literally," or "seriously"?
2. Ask your friends how well you communicate and what your body language says to them. Do you move your hands a lot while speaking? Are your eyes connecting with the audience? By reviewing your style and getting feedback, you can become a better communicator.

18
Organizational Impacts

Allowing people within your organization to be difficult comes at a cost. This cost can be either *tangible* (easily seen) or *intangible* (unseen). When Stan was in the firehouse, people said that they would turn around if they saw his vehicle in the parking lot. That meant fewer responders when the community needed help most. One person created a barrier when we needed staff the most, and that had to change. This chapter breaks down the tangible and intangible costs of keeping toxic individuals around and what leaders and team members can do to take back control of their work environment.

There are seen and unseen impacts when a difficult person is unchecked and working inside the fire station. Workplace conflict costs U.S. companies an estimated $359 billion annually in lost productivity.[1] Difficult people also pose a major threat to organizational health because they become a source of conflict, creating tension that undermines team dynamics and morale; this prompts good people to leave. A Gallup report has showed that employees exposed to chronic negativity experience higher levels of burnout, low engagement, and poor mental health, which all can be directly linked to turnover.[2] Retention is crucial in the fire service, so address the difficult people and improve your seniority list.

Ignoring difficult behaviors allows dysfunction to take root. Toxic emotions at work, often triggered or perpetuated by one or two individuals, can create organizational pain that ripples far beyond the original source.[3] Compassionate leadership and intentional communication are necessary to stop the cycle.

Tangible costs of a difficult person are those that can be seen:

- *Decreased productivity.* Difficult people create distractions and disruptions in the workplace, sapping productivity. Time spent on negative thoughts and engagement could be better used toward positive solutions and ideas.
- *Stress as a result of conflict.* Continued stress leads to mental and physical health problems. Stress leads to increased absenteeism.[4]

- *Recruitment costs.* It is no secret, finding people to serve in the fire service has become difficult and now requires an entire marketing team to entice someone into the fire station. When your organization is suffering from the effects of incivility, bad press, and a toxic culture, you will see a decrease in interest for your open positions. It is all of our jobs to inspire people to join our team by standing up for the culture we need so the lockers are full. People who are subjected to repeated bullying, mistreatment, or incivility have less job satisfaction. If they cannot find support when dealing with a difficult person, they will eventually leave. Replacing them will take time and money, including equipment costs.[5]
- *Damage to reputation.* Difficult people can ruin our reputation simply by association. In a leadership position, difficult behavior can reflect poorly on the organization. Firefighters talk—we know where the trouble is. A bad reputation means fewer applicants for open positions. Furthermore, lawsuits, votes of no confidence, and media attention negatively affect the entire organization—potentially for a long time.
- *Legal costs.* Difficult people cost our organizations money in legal fees and challenge our policies and procedures, consuming our time and efforts. Today, more civil suits are being brought against supervisors than in the past. *You* can be held personally responsible for your lack of leadership or action.

Intangible costs of a difficult person are not immediately visible:

- *Emotional burden.* A difficult person can be emotionally draining and stressful. They can impact the mental health and well-being of others, increasing anxiety, depression, and burnout.
- *Damaged relationships.* A difficult person can damage relationships and make it harder for people to trust one another. They can ruin mutual aid and automatic aid agreements and break down communication and collaboration.
- *Loss of creativity and energy.* A difficult person can create enough conflict that, as a result, firefighters won't share ideas or take risks for fear of attack.
- *Loss of drive and passion.* When there is conflict within the fire station's walls, there will eventually be a loss of drive and passion.
- *Decreased morale and love of the job.* Continued conflict and lack of trust reduce the love of the job. Over time, the stress builds up and takes away the fun.
- *Negative mind-set.* Like a small-minded individual, negativity threatens the organization's forward motion. The good news is your attitude can change, and you become who and what you surround yourself with.

- *Missed opportunities.* Because you are so focused on the people creating issues in your life, you fail to see the opportunities before you.
- *Loss of time and focus in positive areas.* Dealing with people not meeting the mission, vision, core values can be time-consuming and take you away from the positive things in life. Time is money, and the cost of such people is exorbitant.
- *Loss of productivity and focus.* When you're stressed, tired, or burnt out on dealing with these people, you lose productivity and the ability to stay focused on your duties. Stan created a lot of stress in my life. I could not stay focused on what I needed to, so it took me more time to get my day-to-day work done.
- *Loss of respect and professionalism.* When you're constantly challenged, there will be a time when you lose respect for others, and vice versa.
- *Employee retention.* When someone leaves your organization because of a difficult person, you lose a wealth of experience and friendships. Hence, strong retention is crucial.

The organizational costs are overwhelming. If a difficult person is allowed to act freely in your firehouse, it could threaten the history and future of your organization. As a leader, you must act! As firefighters, you need to recognize the threat and work together to mitigate it from the floor. A lot of good can happen when peers work with peers to ensure that their culture remains positive, professional, and free from stress and conflict. Remember that you don't need a particular color of helmet to be a leader. Define the type of culture you want within your team and stand your ground. After all, these are the people coming to help *you* if you get in trouble on the job. We have too many threats already to our health and organization to tolerate dysfunctional behavior.

Chapter 18 Reflection Prompts

1. Write down three examples you have experienced over the past 2 years that have had a negative impact on the organization. How should those have been handled to improve retention, morale, and daily output?
2. What are some ways to recover from negative organizational impacts created by a difficult person?
3. What personal costs does managing people have on you? How do you mitigate those effects?

19
Personal Impacts

Dealing with people's issues—or worse, becoming the target of someone's toxic behavior—can be incredibly stressful. Over time, unaddressed stress can leave lasting marks on both your mental and physical health.[1] I didn't truly grasp the effects that dealing with Stan had on me until afterward. My blood pressure soared, sleep became a rare commodity, and anxiety set in hard. As a new chief, I was tasked with managing a busy fire district; grappling with a bully on top of everything else made my already daunting workload feel almost impossible. I struggled to maintain my focus and composure, my creativity dried up, and I often found myself feeling hopeless. The emotional toll spilled over into my home life. My daughters needed my attention and support when I got home, and my wife relied on me to be present and share quality time together. Instead, the constant stress pulled me away from them. Without the steadfast support of my family and close friends, I honestly don't know how I would have survived that period when I questioned whether I still had what it took to lead.

The personal impacts of dealing with difficult people can be profound, especially when you're dealing with a difficult person on top of everyday pressures. Research has shown that stress triggers your body's fight-or-flight mode, which over time leads to physical issues such as:

- Headaches
- Muscle tension
- Chest pain
- Changes in sex drive
- Problems sleeping

These aren't minor or merely warnings; they're genuine health risks that many of us face when stress goes unchecked.

Beyond the physical effects, stress undermines your mental agility and leadership capabilities. Prolonged exposure to stress can diminish cognitive flexibility and impair decision-making, both of which are essential for effective leadership.[2] When you're stuck in a loop of anxiety and negative thoughts, it's incredibly difficult to stay creative or to see the big picture. That's exactly what happened to me.

The following are signs of stress that I experienced; note that these aren't personal failings but consequences of stress eroding the qualities that make a good leader:

- Loss of focus
- Loss of sleep
- Ruminating (repeatedly thinking about negative thoughts)
- Getting angry over little issues
- Reduced contact with people (friends, family, and colleagues)
- Loss of motivation to exercise
- Decrease in creativity and productivity

Eventually, I learned that regular physical exercise is one of the best antidotes to stress. After my run-in with Stan, I started taking my bicycle to work. I'd ride for miles, letting the rhythmic motion and fresh air clear my mind and restore my sense of balance. This helped me recalibrate, not just physically but mentally and emotionally, too. Our bodies are designed to move, so exercise can help us to break free from the grip of stress and to reset our mood and outlook for the day.

You can find a lot of information online on how to manage your stress and anxiety. Here are a few ways I recommend to manage stress:

- Get regular exercise and eat healthier. A strong body prepares the mind to be strong, too.
- Keep a journal. Write down your concerns, fears, and goals.
- Practice deep breathing and other relaxation techniques, such as yoga. Breathing in and out can lower your heart rate and reset your mood. (I used to laugh at this, but when the alert on my phone reminds me to breathe, I do it.)
- Maintain a sense of humor. Have fun! Clowning around in the office is a blast. Find people who make you laugh and enjoy their time at work.
- Spend time with family and friends. Shut off the noisemakers when with those you care about most.
- Find a hobby and take time to focus on that. Years ago, I started working on vintage automobiles with my dad. Even though he is no longer with us, I love tinkering with Volkswagens buses and dune buggies.

- Listen to music, read a book, garden, write a book, cook, or do any other health-minded activity. Partake in whatever improves your mind-set and energizes your spirit to drive on. (Note there is no mention of drinking alcohol, smoking, or doing drugs, since they will only harm you.)

Difficult Boss

Sometimes it isn't a peer making difficulty in the workplace, but your boss. You're in a tough predicament when it's your boss who is difficult. If you've tried everything—attempting to understand their perspective, discussing your concerns face-to-face, and bridging generational divides—but nothing changes, then ask yourself if this person or environment truly deserves your talent and dedication. Sometimes, the stress inflicted by a boss does more damage than your actual work ever could. If you find that you've become a target and the mistreatment is ongoing, start by following the established chain of command and documenting your concerns. If your immediate supervisor is part of the problem or fails to act, then it becomes appropriate to escalate respectfully, first to their supervisor, then to human resources, or, in cases involving harassment, discrimination, or ethical violations, to legal counsel or the appropriate external authority. Going around the chain of command should never be about avoiding accountability; it should be about protecting yourself and the integrity of the organization. You deserve to work in an environment free from harassment, threats, and harm, both physical and psychological.

Stress isn't just a nuisance; it's a health hazard and a leadership killer. It distorts your priorities, erodes your creativity, and leaves you questioning your abilities. I've been there and learned lessons the hard way. By acknowledging the signs early on, taking steps to mitigate your stress including regular exercise, and leaning on the support of loved ones, you can regain control of your health and leadership. A healthier you will be a better leader, friend, colleague, teammate, and family member.

Remember, our bodies and minds are interconnected. When stress takes over, it doesn't just diminish your performance at work, it seeps into every facet of your life. If you let them, the consequences can be far-reaching. You cannot avoid stress entirely; nevertheless, you can recognize it and manage it proactively, so that it doesn't rob you of your ability to lead effectively or live well. A healthier you will be a better leader.

"I Like Your Shoes"

Years ago, my daughters were working through the negatives of middle school. I wanted them to be able to keep their mind-set positive even in times of negativity, so I told them the following story:

> You found a new pair of shoes at the store that you liked but were expensive. You worked hard to earn money to purchase them and finally brought them home. You immediately put them on and walked through the house. "Nice shoes," I said. The next day, you picked out the best outfit to match your shoes and go to school. At the bus stop, you heard, "Nice shoes." Walking through the halls, you heard, "Nice shoes." You were on cloud nine and feeling great.
>
> Even after all that high praise, the only person you will likely remember commenting on your shoes is the one who says, "Those are the ugliest shoes I have ever seen." That one person may make you never want to wear those shoes again. You miss an abundance of positives when you hang onto the one negative of any day. Such is life.

Move through your day as if everyone loves you and try to understand that some people have a negative mind-set and nothing positive to say. You have a lot to offer the world.

Chapter 19 Reflection Prompts

1. How has work impacted your mental and physical health?
2. Are you using your vacation time to disconnect and relax? If not, why?
3. What hobbies do you have outside work (or has work become your hobby)? What else do you do for yourself outside work?
4. List three things you need to do to improve your health—that, in turn, improve your ability to serve.

20
Putting It All Together

Earlier (see chapters 4 and 6), we defined a difficult person as someone who consistently disrupts an organization's mood, negatively influences others, and disregards your department's mission, vision, and core values. Over time, these individuals stifle the positive energy and progress of your team. They bring unnecessary stress, create retention issues, and hurt recruitment efforts, and they can even expose you to legal challenges. To protect your team and elevate your department's performance, we must all learn to identify these behaviors and work together to counteract the negative energy. Each of us plays a role in maintaining our organization's culture, honor, pride, and reputation. After all, the collective energy of the majority shapes an organization's success, and that energy must be carefully nurtured.

The Lessons I Didn't Have

Nearing the close of this book, you now have insights I didn't possess when I first encountered Stan. Back then, as an operations chief, my focus was on the strategy and technical details (e.g., smooth-bore nozzles vs. automatic), incident command, and the color of smoke. I wasn't prepared to manage someone whose behavior was not only difficult but dangerous, even deviant. Concepts such as narcissism were foreign to me. Yet, after enduring those experiences, I learned that understanding these traits is crucial.

We must go beyond labeling someone as difficult to comprehend the underlying issues that drive their behavior. These might be personal strife, financial problems, or health challenges. Knowledge of the cause(s) of difficult behavior allows you to better support those around you, especially when you might need that person's help in an immediately dangerous to life and health situation.

Do you want a bully, a dangerous individual, or a chronic gossip standing by your side during an emergency? The answer is obvious. You'd rather come into the fire station and experience a positive, uplifting atmosphere—one that fuels teamwork, trust, and resilience.

The Lessons I Wish I'd Had

Looking back, I realize that what I lacked in my own training was a playbook for dealing with difficult people. No one trained me for the human side of leadership. With the lessons in this book, I hope that you will not be caught off guard as I was.

Together, the chapters of this book form the road map I wish I had been given to chart my course through leadership. To make these lessons even more practical, here are some pieces of advice I wish I had been told before I first faced a difficult person:

- *Ignoring entitlement erodes accountability.* If I had known that, I would have set expectations earlier and enforced them more consistently.
- *Gossip spreads like wildfire.* If I had known that, I would have cut it off quickly, instead of letting it fester and cause division.
- *Stress can cost health and clarity.* If I had known that, I would have invested sooner in exercise, balance, and mental well-being.
- *Body language speaks louder than words.* If I had known that, I would have been more conscious of how I carried myself in tough conversations.
- *Organizational culture is fragile.* If I had known that, I would have defended it with the same urgency with which I defended lives on the fireground.
- *Difficult people drain entire teams.* If I had known that, I would have acted faster to protect the good people from being driven away.

These are not just leadership lessons; they are survival tools.

The Challenge and the Charge

By now, you probably recognize that you have already dealt with a difficult person at some point. With this new knowledge, I urge you to identify the challenging traits you see in others. Reflect on generational differences, personal stressors, and external influences contributing to their behavior.

Remember, people become difficult for a variety of reasons: sometimes it's something personal, such as a divorce, financial strain, a major incident, or health issues. I, too, have questioned whether I could have done more to prevent someone from lashing out, yelling, throwing things, or harming me emotionally. It's never easy when your decisions as a leader are unpopular. Yet, those decisions are often made for the greater good of the organization, sometimes even dictated by higher authorities. I never set out to make anyone's life miserable; however, I have been on the receiving end of such treatment, which is a stark reminder that rank comes with responsibility—and the burden of positional power.

If you find yourself the target of one or more difficult individuals, you'll need the courage to stand up for yourself, even if it means standing alone. The personal cost of chronic stress can be high, undermining your health, sleep, and decision-making. Likewise, the very traits that make you a good leader—independence, competence, likeability, and ethics—can also make you a target. Becoming a target may not be about *you* personally, but it can sure feel that way.

The Final Takeaway

If you notice a difficult person in your ranks, pull them aside and listen. Learn their story, reaffirm your policies, and explain how their negativity is affecting the entire organization. If their behavior doesn't improve, you may need to initiate formal disciplinary steps, because keeping the wrong person can cost more than letting them go.

In the end, we will remember not the words of our enemies, but the silence of our friends.[1]

—Dr. Martin Luther King, Jr.

Leadership isn't just about making tough calls; it's about cultivating an environment where every team member feels safe, valued, and respected. The energy you bring has a ripple effect. A single negative force can disrupt an entire team's dynamic, but one strong, positive leader can restore trust, morale, and purpose.

Take a moment each day to ask yourself the following:

- How is the current atmosphere at work affecting my well-being?
- Am I supporting my team in the way they need?
- What can I do today to ensure that everyone remains resilient in the face of adversity?

Closing Thoughts

Thank you for taking the time to read this book and walk through the hard lessons I had to learn without a guide during the most difficult personnel challenge of my career. My sincere hope is that from these lessons you will have gained tools I didn't have, so that you can lead with clarity, and purpose. Today, I strive to lead with my focus on people, protecting those who want to succeed from being derailed by those I fail to manage.

Be safe, have fun, and above all, *don't be difficult!*

Chapter 20 Reflection Prompts

1. Looking back at the difficult people you've encountered, what behaviors stand out as most damaging to your team's culture, and how might you address them differently now?
2. How do you balance holding people accountable and showing compassion when you uncover the stressors or struggles behind their behavior?
3. Which leadership lesson from the book resonates most with you, and how will you put it into practice within the next 30 days?
4. What will you do to protect your own mental and physical health so that you can continue to lead with clarity and resilience?
5. If someone were to describe your leadership style 5 years from now, what would you want them to say, and what steps can you take starting today to make that a reality?

A
Real-World Scenarios

The scenarios here are for you to apply the knowledge you have gained from this book. Consider how you would handle each of them on your own; then think about whom you could go to for help. Also, determine if you have an organizational policy developed to address the topic. If not, what are you going to do to ensure that your organization is prepared to manage these issues?

As you come across difficult people in these scenarios, describe the traits you recognize and determine if their behavior would pose a threat to someone or your organization. Rank the priority of addressing each issue identified.

Scenario 1

Merv constantly comes to the station angry. At the station, he routinely raises his voice. In a recent outburst, he kicked the recycling can over, throwing cans and garbage all over the apparatus bay.

- What trait is Merv displaying?
- How would you handle this incident?
- How could his behavior affect firefighters on shift and the organization?

Scenario 2

Rick claims that the new Genesis tools are worthless, telling everyone that he wants the Hurst extrication equipment instead, because they are lighter. Whenever these are used, Rick makes a scene. In fact, he complains about everything. Several colleagues have come to the office to report they are tired of it.

- What trait is Rick displaying?
- How would you handle this incident?
- How could his behavior affect firefighters and the organization?

Scenario 3

The chief implemented a new policy requiring each firefighter to change hoods after entering a hazardous environment. Captain Johnson disagrees with this new practice and tells his firefighters not to do it because his research states that it does no good in preventing cancer. He would have preferred for the money spent on this to go toward the new European-style helmets he wants. Captain Johnson continuously disregards directives from the chief and other officers.

- What trait is Captain Johnson displaying?
- How would you handle this incident?
- How could his behavior affect firefighters and the organization?

Scenario 4

Mark told the crew that Steve's wife is having an affair with Firefighter Brandon. Brandon is married and has three kids. Mark is always the guy with new information.

- What trait is Mark displaying?
- How would you handle this incident?
- How could his behavior affect firefighters and the organization?

Scenario 5

You're the new chief, visiting each station in your second week on the job. When you visit Station 3, the president of the union comes up to you and states, "Welcome, Chief. I want you to know that the union runs this fire department. If you have any new ideas, please speak to me before implementing them."

- What trait is the union president displaying?
- How would you handle this incident?
- How could his behavior affect firefighters and the organization?

Scenario 6

You are the new battalion chief. Sue, a probationary firefighter, wants to meet with you. During your meeting, she states that the captain is harassing her and she feels uncomfortable. He constantly makes jokes about her homosexuality. When her partner came to the station, he responded by assigning her to do all the station cleaning and extra duties. He does not allow her to train with the other staff.

- What behavior is the captain displaying?
- How would you handle this incident?
- How could his behavior affect firefighters and the organization?

Scenario 7

You're on the fire scene and notice Jamie, the engine operator, stumbling around the truck. Just before the fire call, you saw his vehicle at Duffy's Bar and Grill.

- What trait is Jamie displaying?
- How would you handle this incident?
- How could his behavior affect firefighters and the organization?

Scenario 8

You learn through a Facebook post that the fire engine was used to fill the mayor's swimming pool. When you ask Dante, the firefighter who brought the truck over, about this, he claims that the mayor stopped at the station and ordered him to do it.

- What trait is Dante displaying?
- How would you handle this incident?
- How could his behavior affect firefighters and the organization?

Scenario 9

John went to Home Depot in his uniform. The manager noticed that he was a firefighter and gave him a Milwaukee drill as a gift. John took it home and put it to personal use.

- What trait is John displaying?
- How would you handle this incident?
- How could his behavior affect firefighters and the organization?

Scenario 10

Steve is driving to a car fire. En route to the scene, he blows through an intersection without slowing down, causing vehicles to swerve off the road.

- What trait is Steve displaying?
- How would you handle this incident?
- How could his behavior affect firefighters and the organization?

Scenario 11

Mike states that his captain consistently calls him out at training and tells his peers that he does not want him driving the engine because he had been cited for driving while intoxicated 5 years ago.

- What trait is the captain displaying?
- How would you handle this incident?
- How could his behavior affect firefighters and the organization?

Scenario 12

Gwen states that Will keeps approaching her after calls and asking for a date. She says that it was funny initially but now seems just "creepy."

- What trait is Will displaying?
- How would you handle this incident?
- How could his behavior affects firefighters and the organization?

Scenario 13

The fire department is hiring, and they receive an application from Wade. Before the interview process, one of the lieutenants sees the list of names and tells the Station 1 crew that we should not hire Wade because he is gay.

- What trait is the lieutenant displaying?
- How would you handle this incident?
- How could his behavior affect firefighters and the organization?

Scenario 14

Captain Shaw has been targeting one of the new firefighters because she forgot where a piece of equipment was placed on the truck during a fire scene, which delayed the attack crew from entering the structure for several minutes. Shaw frequently disrespects her, calls her names, and tells others she is "dumb." The new firefighter has become introverted and less upbeat and is ready to quit.

- What trait is Captain Shaw displaying?
- How would you handle this incident?
- How could his behavior affects firefighters and the organization?

Scenario 15

The new recruits arrived at the station today for orientation. While in the classroom, Captain Martinson starts to scream at the candidates, calling them "mindless rookies." He states that they should prepare for a month of hell as they go through training. Martinson excuses this as his way of preparing them to be mentally tough for the job.

- What trait is Captain Martinson displaying?
- How would you handle this incident?
- How could his behavior affect firefighters and the organization?

Scenario 16

Devin complains about a recent change that was made within the records management system. Because of the change, he states that entering reports takes more time now. In his words, this is a "waste of his time," so he is no longer going to enter reports.

- What trait is Devin displaying?
- How would you handle this incident?
- How could his behavior affect firefighters and the organization?

Notes

3. Two Down: The Call That Changed Everything

1. N. Backus, "Murder-Suicide Perpetrator Had Troubled History with Police," *Quad Community Press*, November 3, 2009, https://www.presspubs.com/quad/news/murder-suicide-perpetrator-had-troubled-history-with-police/article_5a5337ff-3bdd-5fb8-9a61-b08a2efb8d46.html.
2. N. Ngo, "Lino Lakes Man Kills Wife, Then Himself, Police Say," *Pioneer Press*, October 2, 2009, https://www.twincities.com/2009/10/02/lino-lakes-man-kills-his-wife-then-himself-police-say/.
3. "Lino Lakes Woman Feared Husband Might Kill Her," *MPR News*, October 3, 2009, https://www.mprnews.org/story/2009/10/03/lino-lakes-woman-feared-husband-might-kill-her.

4. The Different Types of Difficult People

1. C. Maslach, *Burnout: The Cost of Caring* (Cambridge, MA: Malor Books, 2003).
2. P. Lencioni, *The Five Dysfunctions of a Team: A Leadership Fable* (San Francisco: Jossey-Bass, 2002).

5. Why People Matter

1. M. Owen and K. Maurer, *No Easy Day: The Firsthand Account of the Mission That Killed Osama bin Laden* (New York: Dutton, 2012).

6. Dealing with Difficult

1.   I. L. Janis, *Groupthink: Psychological Studies of Policy Decisions and Fiascoes*, 2nd ed. (Boston: Houghton Mifflin, 1982).

2.   Mayo Clinic Staff, *"Chronic Stress Puts Your Health at Risk,"* Mayo Clinic, August 1, 2023, https://www.mayoclinic.org/healthy-lifestyle/stress-management/in-depth/stress/art-20046037.

3.   J. D. Mayer, P. Salovey, and D. R. Caruso, "Emotional Intelligence: New Ability or Eclectic Traits?" *American Psychologist* 63, no. 6 (2008): 503–17, https://doi.org/10.1037/0003-066X.63.6.503.

4.   A. C. McFarlane and R. A. Bryant, "Post-traumatic Stress Disorder in Occupational Settings: Anticipating and Managing the Risk," *Occupational Medicine*, 57, no. 6 (2007): 404–10.

5.   C. R. Rogers, *Client-Centered Therapy: Its Current Practice, Implications, and Theory* (Boston: Houghton Mifflin: 1951).

6.   T. M. Amabile and S. J. Kramer, "The Power of Small Wins," *Harvard Business Review*, May 2011, https://hbr.org/2011/05/the-power-of-small-wins

7. Dealing with Entitlement

1.   *Standard on Fire Department Occupational Safety, Health, and Wellness Program*, NFPA 1500 (Quincy, MA: National Fire Protection Association, 2024).

2.   J. M. Twenge and W. K. Campbell, *The Narcissism Epidemic: Living in the Age of Entitlement* (New York: Free Press, 2009).

3.   Ibid.

4.   Ibid.

5.   W. K. Campbell, A. M. Bonacci, J. Shelton, J. J. Exline, and B. J. Bushman, "Psychological Entitlement: Interpersonal Consequences and Validation of a Self-report Measure," *Journal of Personality Assessment*, 83, no. 1 (2004): 29–45.

6.   A. Bandura, *Social Learning Theory* (Englewood Cliffs, NJ: Prentice Hall, 1977).

8. Dealing with Dangerous

1. NIOSH, "Workplace Violence Prevention Strategies and Research Needs," U.S. Department of Health and Human Services (NIOSH) Publication No. 2006-144, September 2006, https://www.cdc.gov/niosh/docs/2006-144/.
2. American Psychiatric Association, *Diagnostic and Statistical Manual of Mental Disorders*, 5th ed. (Washington, DC: American Psychiatric Association, 2013).
3. National Center for Injury Prevention and Control (U.S.), Division of Violence Prevention, "Preventing Adverse Childhood Experiences (ACEs): Leveraging the Best Available Evidence," Centers for Disease Control and Prevention, U.S. Department of Health and Human Services, 2019, https://stacks.cdc.gov/view/cdc/82316.
4. K. Niven, C. A. Sprigg, and C. J. Armitage, "Emotion Regulation at Work: Understanding the Role of Regulation Strategy Use and Emotional Exhaustion," *European Journal of Work and Organizational Psychology*, 22, no. 5 (2013): 629–45.
5. American Psychiatric Association, *Diagnostic and Statistical Manual*.

9. Dealing with a Narcissist

1. Mayo Clinic Staff, "Narcissistic Personality Disorder," April 6, 2023, https://www.mayoclinic.org/diseases-conditions/narcissistic-personality-disorder/symptoms-causes/syc-20366662.
2. S. Thomaes, B. J. Bushman, B. O. de Castro, and H. Stegge, "What Makes Narcissists Bloom? A Framework for Research on the Etiology and Development of Narcissism," *Child Development Perspectives*, 3, no. 1 (2009): 30–35.
3. T. Millon, S. Grossman, S. Meagher, C. Millon, and R. Ramnath, *Personality Disorders in Modern Life*, 2nd ed. (Hoboken, NJ: Wiley, 2004).
4. E. Ronningstam, *Identifying and Understanding the Narcissistic Personality* (New York: Oxford University Press, 2005).

5. W. K. Campbell and C. A. Foster, "The Narcissistic Self: Background, an Extended Agency Model, and Ongoing Controversies," in *The Self*, ed. C. Sedikides and S. J. Spencer (New York: Psychology Press, 2007), 115–38.

6. American Psychiatric Association, *Diagnostic and Statistical Manual of Mental Disorders*, 5th ed. (Washington, DC: American Psychiatric Association, 2013).

10. Dealing with Defamation

1. American Law Institute, Restatement (Second) of Torts §558 (1977).

2. Pickering v. Board of Education, 391 U.S. 563 (1968).

3. G. Namie, "2021 WBI U.S. Workplace Bullying Survey," Workplace Bullying Institute, 2021, https://workplacebullying.org/wp-content/uploads/2021/04/2021-Full-Report.pdf.

4. Ibid.

5. National Society of Executive Fire Officers, "Firefighter Code of Ethics," International Fire Chiefs Association, accessed November 5, 2025, https://www.iafc.org/docs/default-source/1efo/firefighter-code-of-ethics.pdf.

6. Pickering v. Board of Education.

7. *Standard on Fire Department Occupational Safety, Health, and Wellness Program*, NFPA 1500 (Quincy, MA: National Fire Protection Association, 2024).

8. American Psychological Association, *2024 Work in America™ Survey: Psychological Safety in the Changing Workplace*, June 2024, https://www.apa.org/pubs/reports/work-in-america/2024/2024-work-in-america-report.pdf.

9. Ibid.

11. Dealing with a Workplace Bully

1. I encourage you to take this online survey, which is available at https://www.surveymonkey.com/r/FireServicebully. You will see results after completing the survey.

2. Workplace Bullying Institute. "Definition of Workplace Bullying." Workplace Bullying Institute, 2021. https://workplacebullying.org/individuals/problem/definition/.

3. M. B. Nielsen and S. Einarsen, "Outcomes of Exposure to Workplace Bullying: A Meta-analytic Review," *Work & Stress*, 26, no. 4 (2012): 309–32.
4. G. Namie, "2021 WBI U.S. Workplace Bullying Survey," Workplace Bullying Institute, 2021, https://workplacebullying.org/wp-content/uploads/2021/04/2021-Full-Report.pdf.
5. Workplace Bullying Institute, "2024 WBI U.S. Workplace Bullying Survey," Workplace Bullying Institute, 2025, https://workplacebullying.org/2024-wbi-us-survey/.
6. K. Cherry, "What Is Conformity?" *Verywell Mind*, July 28, 2023, https://www.verywellmind.com/what-is-conformity-2795889.
7. Ibid.
8. G. Graham, *Risk Management for Public Safety Agencies* (Lexipol / California POST Command College, 1996). *(Commonly cited for the phrase "If it is predictable, it is preventable.")*

12. Dealing with Disgruntled

1. R. J. Bies and T. M. Tripp, "Beyond Distrust: 'Getting Even' and the Need for Revenge," in *Trust in Organizations: Frontiers of Theory and Research*, ed. R. M. Kramer and T. R. Tyler (London: Sage, 1996), 246–60.
2. J. W. Kassing, "Consider This: A Comparison of Factors Contributing to Employees' Expressions of Dissent," *Communication Quarterly*, 56, no. 3 (2008): 342–55.
3. C. Maslach and M. P. Leiter, "Burnout: A Multidimensional Perspective," in *The Fulfilling Workplace: The Organization's Role in Achieving Individual and Organizational Health*, ed. R. J. Burke (London: Routledge, 2013), 51–70.

13. Dealing with Small-Minded People

1. K. E. Stanovich, *What Intelligence Tests Miss: The Psychology of Rational Thought* (New Haven, CT: Yale University Press, 2009).
2. L. Festinger, *A Theory of Cognitive Dissonance* (Evanston, IL: Row, Peterson, 1957).
3. J. P. Kotter, *Leading Change* (Boston: Harvard Business Review Press, 2012).

4. Stanovich, *What Intelligence Tests Miss.*
5. Festinger, *A Theory of Cognitive Dissonance.*

14. Dealing with Deviance

1. S. P. Robbins and T. A. Judge, *Organizational Behavior,* 18th ed. (New York: Pearson, 2021).
2. *Standard on Fire Department Occupational Safety, Health, and Wellness Program,* NFPA 1500 (Quincy, MA: National Fire Protection Association, 2024).
3. S. P. Robbins and T. A. Judge, *Organizational Behavior.*
4. Federal Emergency Management Agency. *Leadership and Supervision (Student Manual).* Emmitsburg, MD: National Fire Academy, U.S. Department of Homeland Security, 2017, https://usfa.fema.gov/downloads/pdf/nfa/student-manual-leadership-and-supervision.pdf.
5. *Standard on Fire Department Occupational Safety, Health, and Wellness Program,* NFPA 1500 (Quincy, MA: National Fire Protection Association, 2021).

15. Dealing with a Gossip

1. G. Michelson and S. Mouly, "Rumour and Gossip in Organisations: A Conceptual Study," *Management Decision,* 38, no. 5 (2000): 339–46.
2. R. Gassaway, "Situational Awareness Matters," accessed October 22, 2025, https://www.samatters.com.
3. R. I. M. Dunbar, "Gossip in Evolutionary Perspective," *Review of General Psychology,* 8, no. 2 (2004): 100–110.
4. N. B. Kurland and L. H. Pelled, "Passing the Word: Toward a Model of Gossip and Power in the Workplace," *Academy of Management Review,* 25, no. 2 (2000): 428–38.
5. *Standard on Fire Department Occupational Safety, Health, and Wellness Program,* NFPA 1500 (Quincy, MA: National Fire Protection Association, 2020).

16. Generational Differences

1. Simple Living Global, "Social Media," https://simplelivingglobal.com/social-media/.
2. Beaven, J. (2014), "Generational Differences in the Workplace: Thinking Outside the Boxes," *Contemporary Journal of Anthropology and Sociology*, 4: 68–80.
3. M. Richtel and A. Flanagan, "It's Life or Death: The Mental Health Crisis Among U.S. Teens," *New York Times*, July 31, 2022, https://www.nytimes.com/2022/07/31/health/mental-health-crisis-teens.html.

17. Understanding Body Language

1. A. Mehrabian, *Silent Messages* (Belmont, CA: Wadsworth Publishing, 1971).
2. M. S. Rao, "Tools and Techniques to Boost the Eloquence of Your Body Language in Public Speaking," *Industrial and Commercial Training* 49, no. 2 (2017): 75–79, https://doi.org/10.1108/ICT-08-2016-0053.

18. Organizational Impacts

1. "Workplace Conflict and How Businesses Can Harness It to Thrive: CPP Global Human Capital Report," CPP, July 2008, https://www.jamspathways.com/hubfs/CPP_Global_Human_Capital_Report_Workplace_Conflict.pdf.
2. "State of the Global Workplace: 2022 Report," Gallup, 2022, https://millennium-challenge.com/wp-content/uploads/2023/01/state-of-the-global-workplace-2022-download-1.pdf.
3. P. J. Frost, *Toxic Emotions at Work: How Compassionate Managers Handle Pain and Conflict* (Boston: Harvard Business School Press, 2003.
4. W. Darr and G. Johns, "Work Strain, Health, and Absenteeism: A Meta-analysis," *Journal of Occupational Health Psychology*, 13, no. 4 (2008): 293–318.

5. L. M. Cortina, "Unseen Injustice: Incivility as Modern Discrimination in Organizations," *Academy of Management Review*, 33, no. 1 (2008): 55–75.

19. Personal Impacts

1. Mayo Clinic Staff, "Stress Symptoms: Effects on Your Body and Behavior," Mayo Clinic, accessed October 19, 2025, https://www.mayoclinic.org/healthy-lifestyle/stress-management/in-depth/stress-symptoms/art-20050987.

2. American Psychological Association, "Stress Effects on the Body," *American Psychological Association*, created November 1, 2018; last reviewed October 21, 2024, https://www.apa.org/topics/stress/body.

20. Putting It All Together

1. M. L. King, Jr., *The Trumpet of Conscience* (New York: Harper & Row, 1968).

Index

A

absenteeism and stress 115
abuse 47
accountability (lack of) 43, 124
   dangerous individuals and 46
   defamation and 59
   deviance and 84
   resistance to 41
action, immediate and deviance 86
active listening 35–36
addressing bullying 67–68
   build a culture of psychological
     safety 69
   restorative justice 68
   set the tone 68
   support targets of bullying 69
   take immediate action 69–70
addressing conflicts 75
addressing dangerous individuals 49–50
   boundaries (setting) 49
   employee assistance programs
     (EAPs) 49
   professional mental health services 49
   self-care and 49
   support network for 49
addressing defamation 60–61
   acknowledge the behavior
     immediately 60
   document everything 60
   educate and prevent 61
   enforce policy and discipline fairly 61
   ensure psychological safety 61
   follow the chain of command 60
   leaders modeling the standard 61
addressing deviance 86
   develop clear and concise policies and
     procedures 86
   follow up 86
   help employees to feel supported 86
   immediate action 86

addressing difficult behavior 34–35
   documentation 35
   network of trusted peers 35
   practice active listening 35–36
   setting boundaries 34–35
   staying calm and professional 34–35
addressing disgruntled employees 74–75
   address conflicts 75
   creating a positive work
     environment 75
   listening 74
   providing support 75
addressing entitlement 43
   balance praise with responsibility 43
   encourage empathy and humility 43
   lead by example 43
   promote a culture of service over
     status 43
   set clear expectations and
     accountability 43
addressing generational
     differences 103–104
   adjust communication styles 103
   find strengths instead of fighting
     differences 104
   modernize training approaches 104
   set clear expectations 104
addressing gossip 92–93
   address rumors swiftly 92
   discourage clique formation 92
   educate firefighters 92
   encourage open communication 92
   take disciplinary action when
     necessary 92
addressing narcissism 54
   avoid drama 54
   boundaries (setting) 54
   documenting 55
   do not enable 54
   professional mental health services 54

addressing small-mindedness 80
    encourage collaboration 81
    encourage open-mindedness 80
    lead by example 81
    provide training and development
        opportunities 81
admiration 52
adverse childhood experiences 47
advise for dealing with difficult
        behavior 124
    body language speaks louder than
        words 124
    difficult people drain entire
        teams 124
    gossip spreads like wildfire 124
    ignoring entitlement erodes
        accountability 124
    organizational culture is fragile 124
    stress can cost health and clarity 124
aggression 46
America Online 100
anger and defamation 59
antisocial personality disorder 48
anxiety 119
    managing 120
Army National Guard 5
attention
    giving one your full 108
    seeking and defamation 59
authority, resistance to 32
avoiding conflict 21

B
baby boomers (boomers) 99
behavior
    toxic 21
belonging and gossip 91
bin Laden, Osama 27
biological factors of narcissism 54
bipolar disorder 48
BlackBerrys 100
blame shifting and disgruntled
        employees 72
blended learning 104
blood-borne pathogens 77
body language 107–112, 124
    attention 108
    context matters 109
    cultural factors 109
    eye contact 108
    facial expressions 109
    features of 108–112
    gender 109

generational differences 109
    gestures 108
    posture 108
    video-record yourself 108
boomers 99
borderline personality disorder 48
bosses and difficult behavior 121
boundaries (setting) 49
    addressing narcissism 54
bullying 20, 63–67
    definition of 65–66
    emotional 66
    health consequences of 66
    mobbing 65
    offensive nonverbal conduct 66
    physical 66
    social media and 66
    support targets of 69
    synonyms for 66
    technological 66
    verbal 66
    withholding information 66
    in the workplace 63
burnout and disgruntled employees 74

C
camaraderie 45
causes of bullying 66
    command 67
    fear of retaliation 67
    hierarchy 67
    team dynamics 67
    traditions 66
causes of dangerous individuals 47–48
    abuse 47
    adverse childhood experiences 47
    antisocial personality disorder 48
    bipolar disorder 48
    borderline personality disorder 48
    childhood trauma 47
    cultural enablers 48
    fear 48
    insecurity 47
    isolation 48
    lack of emotional regulation 47
    low self-esteem 47
    mental health conditions 48
    organizational enablers 48
    post-traumatic stress disorder
        (PTSD) 48
    shame 48
causes of defamation 59–60
    anger 59

attention seeking 59
communication gaps 59
ego 59
generational differences 59
lack of accountability 59
poor leadership 59
resentment 59
technological differences 59
causes of deviance 85
frustration with leadership 85
perceived injustice 85
causes of difficult behavior 33–34
lack of emotional awareness 33
lack of skills 33
organizational misinformation 33
past workplace trauma 34
personal struggles 33
rumors 33
stress 33
causes of disgruntled employees 73–74
burnout 74
disciplinary action 75
job stress 74
lack of recognition 73
perceived injustice 73
poor leadership 73
proof communication 73
causes of entitlement 41–42
cultural reinforcement within the
department 42
length of service without
accountability 42
misinterpreted praise and
recognition 42
causes of gossip 91–92
communication gaps 91
competition 91
confabulation 91
disengagement 91
lack of information 91
low morale 91
need for belonging 91, 91–92
social connection 91, 91–92
unaddressed conflict 91
causes of narcissism 53
biological factors 54
early childhood environment 53
fragile ego 53
low self-esteem 53
psychological factors 54
social reinforcement 53
causes of small-mindedness 79–80
cognitive dissonance avoidance 80

cognitive rigidity 80–81
cultural insulation 80
social conditioning 80
Certificate of Achievement 4
chain of command and addressing
defamation 60
change and resistance to from
small-minded individuals 78
childhood trauma 47
cliques 90
gossip and 92
cognitive dissonance avoidance and
small-mindedness 80
cognitive rigidity and
small-mindedness 80–81
collaboration and small-mindedness 81
Columbine High School 100
combat 3
medics 1
command and bullying 67
communication
adjusting styles to address
generational differences 103
assimilation 109
chain 36
distortion 36
gaps and causes of defamation 59
gaps and gossip 91
intentional 115
open and gossip 92
poor and causes of disgruntled
employees 73
competition and gossip 91
confabulation 91
conflict
addressing for disgruntled
employees 75
avoiding 21
unaddressed and gossip 91
consistently negative 32
creating alliances and gossip 90
creativity, loss of due to difficult
behavior 116
critical decision-making 31
cultural dynamics and deviance 86
cultural enablers 48
cultural factors and body language 109
cultural insulation and
small-mindedness 80

D

daily negativity and disgruntled
employees 73

damaged relationships and difficult
    behavior 116
dangerous individuals 20, 45–48
    accountability and 46
    aggression and 46
    causes of 47–48
    common characteristics of 46–47
    defining dangeous 46–47
    lack of empathy and 46
    manipulative behavior and 46
    violence and 46
decision-making (critical) 31–34
defamation 20, 57–60
    criteria for 58–59
    definition of 58–59
    free speech vs. 61
    libel 58
    slander 58
Delayed Entry Program 2
development opportunities and
    small-mindedness 81
deviance 20, 83–85
    characteristics of 84
    clear policy enforcement for 85
    cultural dynamics of 86
    definition of 84–85
    disregard for rules and policies 84
    disruptive or harmful behavior 84
    erosion of ethical standards 85
    external pressures 85
    lack of accountability 84, 85
    misalignment with organizational
        values 85
    negative peer influence 84
    peer influence and 84
    personal stress and 85
    team dynamics and 86
*Diagnostic and Statistical Manual of
    Mental Disorders* 54
difficult behavior 19–22
    addressing 34–35
    behaviors to watch out for 21
    from your boss 121
    bullying 20
    causes of 33–34
    common characteristics of 32
    consistently negative 32
    dangerous 20
    dealing with 31–34
    defaming 20
    definition of 20, 32, 123
    deviance 20
    disgruntled 20

disruptive to team cohesion 32
    emotionally volatile 32
    entitled 20
    in the fire service 19
    gossiping 20
    leadership and 125–126
    narcissistic 20
    organizational impacts of 115–116
    personal impact of dealing
        with 119–121
    prolonged exposure to 20
    recognizing 21
    resistant to authority and policy 32
    small-minded 20
    types of 20
    understanding 123
digital natives 101
disciplinary actions 22
    addressing defamation 61
    disgruntled employees and 75
    gossip and 92
disengagement and gossip 91
disgruntled employees 20, 71–74
    blame shifting and excuses 72
    common characteristics of 72
    daily negativity and 73
    definition of 72–73
    distrust in leadership 72
    low engagement 72
    persistent criticism without solutions
        from 72
    resistance to change 72
    withdrawal and low engagement 72
dispatchers 6
disruptive behavior
    deviance and 84
    teams and 32
    training and 124
distrust in leadership and disgruntled
        employees 72
documentation
    addressing defamation 60
    addressing narcissism 55
    dealing with difficult individuals 35
drama (engaging in) and addressing
        narcissism 54
drill instructors 22
drive, loss of due to difficult
        behavior 116
dysfunctional teams, model of 21

E

EAPs (employee assistance programs) 49

early childhood environment and
        narcissism 53
education
    defamation and 61
    of firefighters about gossip 92
ego
    causes of defamation 59
    fragile 53
    teamwork vs 42
elite professionals 27
emotions
    awareness of 33
    bullying and 66
    burden of and intangible impacts of
        difficult behavior 116
    regulation of 47
    volatility of 32
empathy 43, 52
    lack of 46
employee assistance programs (EAPs) 49
employees
    protecting 124
    support from deviant behavior 86
enabling
    addressing narcissism 54
    organizations and 48
energy, loss of due to difficult
        behavior 116
engagement, low and disgruntled
        employees 72
entitlement 20, 39–42
    belief that rules do not apply to
        them 41
    characteristics of 41
    definition of 40–41
    environment and 42
    expectation of special treatment 41
    narcissism and 40, 52
    resistance to accountability 41
    safety and 40–41
    self-importance and 40
environment and entitlement 42
equipment 26
ethical standards, erosion of and
        deviance 85
evolution 1
excuses and disgruntled employees 72
expectations 43
    generational differences and 104
Expert Marksmanship rank 4
external pressures and deviance 85
eye contact 108

F
Facebook 100
facial expressions 109
fear 48
    of retaliation and bullying 67
fight-or-flight mode 119
fire chiefs 25
fire departments
    characteristics of 28
    equipment for 26
    integrating generations in 102
    staffing 26
    starting a new department 25
    technicians 26
    violence and 46
    weapons in 49
firefighters
    career 6
    education and gossip 92
    as elite professionals 27
    full-time 6
    placed on a pedestal 19
    protecting 124
    resume for becoming 1
    tradition and 19
    training to become 5
fire service
    discipline and 22
    military vs 6
    shared mission and 22
    trust and 22
focus, loss of due to difficult
        behavior 117
follow up and deviance 86
Fort Bliss, TX 2
free speech vs. defamation 61

G
generational differences 95–101
    baby boomers (boomers) 99
    body language and 109
    causes of defamation 59
    definitions of 96–99
    Generation Alpha 102
    Generation X (latchkey
        generation) 99
    Generation Y (millennials) 100
    Generation Z (iGen) 101
    integrating into the fire
        departments 102
    Silent Generation 98
    summary table of 97–98, 98–99
Generation Alpha 102

Generation X (latchkey generation) 99
Generation Y (millennials) 100
Generation Z (iGen) 101
gestures 108
Glasgow, Sergeant First Class 2
globalization 100
gossip 20, 37, 89–91
    creating alliances 90
    definition of 90–91
    engaging in negative talk 90
    sharing rumors or private
        information 90
    spending excessive time
        socializing 90
    spreading 124
grandiosity 52
Great Depression 97
group alliances 90
groupthink 31

H
hand grenade training 3–4
harmful behavior and deviance 84
Hennepin County Water Patrol
    Division 6
hierarchy and bullying 67
humanistic therapy 35
humility 43

I
inflexible thinking and small-minded
    individuals 79
information, lack of and gossip 91
injustice and deviance 85
insecurity 47
    small-mindedness and 79
intangible organizational impacts of
    difficult behavior 115
    damaged relationships 116
    decreased morale and love of the
        job 116
    emotional burden 116
    employee retention 117
    loss of creativity and energy 116
    loss of drive and passion 116
    loss of productivity and focus 117
    loss of respect and
        professionalism 117
    loss of time and focus in positive
        areas 117
    missed opportunities 117
    negative mind-set 116
intent and impact and defamation 61

irrelevance and small-mindedness 79
isolation 48

J
job stress and disgruntled employees 74
John, Elton 99
judgment 48
    small-minded individuals and 79

K
keyboard commander 59
Kotter, John 79

L
lack of emotional awareness 33
lack of recognition and disgruntled
    employees 73
lack of skills 33
latchkey generation 99
leadership
    addressing defamation 61
    avoiding conflict and 21
    compassionate 115
    difficult behavior and 125–126
    distrust in from disgruntled
        employees 72
    by example 43, 81
    frustration with and deviance 85
    lessons of 1–4
    military 3, 5
    poor and causes of disgruntled
        employees 73
    professionalism and 61
learning
    blended 104
    vertical curve 6
legal costs and tangible organizational
    impacts of difficult behavior 116
Lencioni, Paul 21
length of service without
    accountability 42
libel 58
listening
    active 35
    addressing disgruntled employees 74
    reflective model 35–36
loss of time and focus in positive areas
    and intangible organizational
        impacts of difficult behavior 117
love of the job, decreased due to difficult
    behavior 116
low engagement and disgruntled
    employees 72

low morale and gossip 91
low self-esteem 47

M

M16A1 automatic rifle training 3–4
    Expert Marksmanship rank 4
manipulative behavior 46, 52
Mayday 29
mental health conditions 48
Mercury, Freddie 99
military 1
    Certificate of Achievement 4
    Delayed Entry Program 1
    drill instructors 22
    Expert Marksmanship rank 4
    fire service vs 6
    leadership 5
millennials 100
mirroring emotions 35
misinformation 33
misinterpreted praise and recognition 42
missed opportunities and intangible organizational impacts of difficult behavior 117
mobbing 65
morale 41
    decreased due to difficult behavior 116

N

Namie, Gary 52
Namie, Ruth 52
narcissism 20, 40, 51–53
    common characteristics of 52, 52–53
    definition of 52–53
    early childhood environment 53
    entitlement 52
    fragile ego 53
    grandiosity 52
    lack of empathy 52
    manipulative behavior 52
    need for admiration 52
    self-importance 52
narcissistic personality disorder 54
National Fire Protection Association (NFPA) 26
    NFPA 1500 39
National Institute of Occupational Safety and Health (NIOSH) 45
    recognizing warning signs of escalation 45
National Institute of Standards and Technology 26

Navy SEALs 27
negative mind-set
    intangible organizational impacts of difficult behavior and 116
    positive mind-set to offset 122
negativity
    disgruntled employees and 73
    gossip and 90
network of peers 35
new experiences (closed off to) and small-minded individuals 79
9-1-1 28
No Easy Day 27

O

Obama, Barack 27
Occupational Safety and Health Administration (OSHA) 77
offensive nonverbal conduct bullying 66
open-mindedness vs. small-mindedness 80
organizations
    enablers 48
    fagility of culture 124
    impacts of difficult behavior 115–116
    misinformation in 33
    values and misalignment (deviance) 85
OSHA (Occupational Safety and Health Administration) 77
Owen, Mark 27

P

Pan Am Flight 103 100
paraphrasing 35
passion, loss of due to difficult behavior 116
past workplace trauma 34
peers
    negative influence of and deviance 84–85
    network of trusted 35
people matter 28
perceived injustice and disgruntled employees 73
performance 41
persistent criticism without solutions and disgruntled employees 72
personal impacts of difficult behavior 119–121
    physical issues 119–121
personal stress and deviance 85
personal struggles 33

physical bullying 66
physical exercise 120
    stress and 120
Pickering v. Board of Education 58
policies
    disregard for and deviance 84
    enforcement and addressing
        defamation 61
    enforcement and deviance 85
    procedures and deviance 86
    resistant to 32
poor leadership and defamation 59
positive mind-set vs. negative
        mind-set 122
post-traumatic stress disorder
        (PTSD) 20, 48
posture 108
practice active listening 35–36
praise 43
    misinterpreted 42
    recognition and 42
    unity and 42
prevention of defamation 61
productivity
    decreased due to difficult
        behavior 115
    loss of due to difficult behavior 117
professionalism 34–35, 61
    loss of due to difficult behavior 117
professional mental health services 49
    addressing narcissism 54
providing support and disgruntled
        employees 75
psychological factors and narcissism 54
psychological safety
    bullying and 69
    defamation and 61
PTSD (post-traumatic stress disorder) 20, 48
public safety 58

R
recognition 42
    lack of and disgruntled employees 73
    misinterpreted 42
recruitment, costs of due to difficult
        behavior 116
reflective listening model 35
reputation, damage to from difficult
        behavior 116
resentment and defamation 59
resistance to change
    disgruntled employees and 72
    small-minded individuals and 78

resistant to authority and policy 32
respect, loss of due to difficult
        behavior 117
responsibility 43
restorative justice 68
retention 115
    intangible organizational impacts of
        difficult behavior and 117
Rogers, Carl 35
Roseville, MN 5
rules, disregard for and deviance 84
rumors 33
rumors and gossip 92

S
safety
    entitlement and 40–41
    maintaining a culture of 39
self-care 49
self-esteem (low) 47, 53
self-importance 40, 52
service vs. status 43
setting boundaries 34–35
shame 48
shared mission 22
Silent Generation 98
skills (lack of) 33
slander 58
small-mindedness 20, 77–79
    characteristics of 78
    closed off to new experiences 79
    definition of 78–79
    fear of insecurity 79
    fear of irrelevance 79
    inflexible thinking 79
    judgmental 79
    resistance to change 78
Snapchat 100
social conditioning and
        small-mindedness 80
social connection and gossip 91
socializing and gossip 90
social learning theory 42
social media 59
    bullying and 66
social reinforcement and narcissism 53
staffing 26
status vs. service 43
staying calm 34–35
stress 33
    absenteeism and 115
    anxiety and 119
    clarity and 124

costs of health and 124
   disgruntled employees and 74
   fight-or-flight mode and 119
   managing 120
   physical exercise and 120
   signs of 120–121
   tangible organizational impacts of
      difficult behavior 115
   unaddressed 119
summarizing 35
support network 49
Supreme Court of the United States 58

T
tangible organizational impacts of
      difficult behavior 115
   damage to reputation 116
   decreased productivity 115
   legal costs 116
   recruitment costs 116
   stress as a result of conflict 115
teams
   bullying and 67
   cohesion, disruptions to 32
   deviance and 86
   ego vs 42
   model of dysfunctional 21
technicians 26
technological bullying 66
technological differences 78
   causes of defamation 59
toxic behavior 21
tradition 19
   bullying and 66
   tolerance and 67
training 3
   generational differences and 104
   small-mindedness and 81

Trans World Airlines Flight 840 100
trauma, childhood 47
trust 22, 41

U
unity and praise 42
U.S. Army 5

V
verbal bullying 66
vertical learning curve 6
violence 46
   tolerating 46

W
weapons 49
White Sands Desert 3
wisdom 1
withdrawal and disgruntled
      employees 72
withholding information bullying 66
wolf packs 90
workplace
   conflict, fiscal cost of 115
   harmony 58
   past trauma in 34
   positive for addressing disgruntled
      employees 75
   retention 115
workplace bullying 63–67
   definition of 65
   illegality of 68
Workplace Bullying Institute 52, 65, 67
World Trade Center attack 99
World War II 98

Y
Yahoo 100